LIVESTOCK PRODUCTION AND MANAGEMENT

LIVESTOCK PRODUCTION AND MANAGEMENT

Dr. S.K. Singh

RANDOM PUBLICATIONS

NEW DELHI - 110 002 (INDIA)

Livestock Production and Management

ISBN 978-93-51116-87-5

Published in 2015 in India by

RANDOM PUBLICATIONS

4376-A/4B, Gali Murari Lal, Ansari Road
New Delhi-110 002
Phone: +9111-43580356, 23289044
E-mail: randomexports@gmail.com; sales@randompublications.com; info@randompublications.com
Reprinted 2019

Type Setting by: Friends Media, Delhi-110089
Digitally Printed at: Replika Press Pvt. Ltd.

Preface

Livestock are domesticated animals raised in an agricultural setting to produce commodities such as food, fibre and labour. This article does not discuss poultry or farmed fish, although these, especially poultry, are commonly included within the meaning of "livestock". Livestock are generally raised for profit. Raising animals is a component of modern agriculture. It has been practiced in many cultures since the transition to farming from hunter-gather lifestyles.

Animal-rearing has originated during the cultural transition to settled farming communities rather than hunter-gatherer lifestyles. Animals are 'domesticated' when their breeding and living conditions are controlled by humans. Over time, the collective behaviour, life cycle, and physiology of livestock have changed radically. Many modern farm animals are unsuited to life in the wild. Dogs were domesticated in East Asia about 15,000 years ago, Goats and sheep were domesticated around 8000 BC in Asia. Swine or pigs were domesticated by 7000 BC in the Middle East and China. The earliest evidence of horse domestication dates to around 4000 BC. The term "livestock" is nebulous and may be defined narrowly or broadly. On a broader view, livestock refers to any breed or population of animal kept by humans for a useful, commercial purpose. This can mean domestic animals, semi-domestic animals, or captive wild animals. Semi-domesticated refers to animals which are only lightly domesticated or of disputed status. These populations may also be in the process of domestication. Some people may use the term livestock to refer to only domestic animals or even to only red meat animals. During the history of animal husbandry, many secondary products have arisen in an attempt to increase carcass utilization and reduce waste. Farming practices vary dramatically worldwide and between types of animals. Livestock are generally kept in an enclosure, are fed by human-provided food and are intentionally bred, but some livestock are not enclosed, or are fed by access to natural foods, or are allowed to breed freely, or any combination thereof.

Livestock raising historically was part of a nomadic or pastoral form of material culture. The herding of camels and reindeer in some parts of the world remains unassociated with sedentary agriculture. Livestock diseases compromise animal welfare, reduce productivity, and can infect humans. Animal diseases may be tolerated, reduced through animal husbandry, or reduced through antibiotics and vaccines. In developing countries, animal diseases are tolerated in animal husbandry, resulting in considerably reduced productivity, especially given the low health-status of many developing country herds. Disease management for gains in productivity is often the first step taken in implementing an agriculture policy. Disease management can be achieved through changes in animal husbandry. These measures may aim to control spread using biosecurity measures, such as controlling animal mixing, controlling entry to farm lots and the use of protective clothing, and quarantining sick animals. Diseases also may be controlled by the use of vaccines and antibiotics. Since many livestock are herd animals, they were historically driven to market "on the hoof" to a town or other central location. Livestock production constitutes a very important component of the agricultural economy of developing countries, a contribution that goes beyond direct food production to include multipurpose uses, such as skins, fibre, fertilizer and fuel, as well as capital accumulation. Furthermore, livestock are closely linked to the social and cultural lives of several million resource-poor farmers for whom animal ownership ensures varying degrees of sustainable farming and economic stability. There is a growing trend among farmers to support welfare-friendly and sustainable production in livestock farming, with milk produced by cows that graze in fields, free-range meat and eggs, and organic produce. Animals are provided with more space, outdoor runs with shelter and daylight in their housing. These measures go beyond legal standards for animal welfare. Appropriate sustainable livestock management practices are required so that livestock keepers can take advantage of the increasing demand for livestock products and protect their livestock assets in the face of changing and increasingly variable climates.

This book is written for all those involved in animal husbandry as a guide into the production techniques involved in livestock and poultry.

I thank all members of my team who have helped in the preparation of the book. My special thanks go to "Random Publications" who have published the book.

— *Dr. S.K. Singh*

Contents

Chapter 1

Issues and Perspectives in Livestock Production

The MRH system applies to approximately 14 percent of the global population. This ratio is particularly high in sub-Saharan Africa where 41 percent of the region's population is associated with the system and in Central and South America where it is 35 percent. The system is replacing grazing systems in Africa and Latin America. In Africa, the process is mainly driven by population growth, and, in Central and South America, by economic development and technological innovations. In many parts of the world, farmers are clearing rainforests to expand this system e.g. in South America along the Andean foothills (the western border of Amazonia) and in Central America. In Africa, this process is somewhat constrained by the tse-tse/trypanosomiasis complex.

This system is particularly important for large areas of sub-Saharan Africa. The main challenge is finding ways to increase the productivity of the system under serious constraints for both public and private investments. It is generally acknowledged that the biological potentials of mixed systems will be the key to productivity increases, and the expectation is that purchased feed inputs will be replaced by enhanced knowledge about the system, in particular, nutrient cycled within the system. In the more humid parts of Asia, annual crops have been replaced by perennial crops and livestock play a minor role.

In Latin America, low population density, high degrees of urbanization and relatively high *per caput* incomes have induced farming systems generally more oriented towards livestock production. In the tropical rain-forest regions, very resource consuming systems were established, in some cases driven by policies and in others by poverty. Many of the policies that promoted wasteful utilization of these resources have been stopped in the process of structural adjustment.

Arid and Semi-arid Tropics and Sub-tropics (MRA)

The MRA system is a mixed farming systems in tropical and subtropical regions with a vegetation growth period of less than 180 days. The main restriction of this system is the low primary productivity of the land due to low rainfall. The more severe the constraint, the less important crops become in the system and the more livestock take over as the primary income and subsistence source. The system is important in the West Asia and North Africa region, in parts of the Sahel (Burkina Faso, Nigeria), in large parts of India, and less important in Central and South America. Typical cases are dryland farming-sheep systems in northern Africa and in the Indian subcontinent, and also the small ruminant-cassava systems in northeastern Brazil.

Resources and Production

The more arid the conditions become the greater the necessity to keep livestock as an asset for farmers. Given the low intensity of the system and the multiple purposes of livestock, the introduction of improved breeds has been quite limited. Thus, loss of domestic animal biodiversity is not likely to be very significant under these conditions.

Globally, 11 percent of the world cattle population and 14 percent of sheep and goats are found in this system. Small ruminants are particularly important in West Asia and North Africa under the MRA system. Grazing land not suited for crop production is the main feed resource of the system supported by strategic use of crop stubbles and straw. Land not used for cropping is frequently community owned. Traditional rules on access to this resource have frequently not withstood the changes occurring in the last decades, particularly population pressure. This leads to common problems of overgrazing and resource degradation.

Given the high risk involved in crop production this system tends to produce crops mainly for subsistence. They are usually produced very extensively, thus minimizing the financial risks but also limiting the potential for good harvests. Livestock are produced extensively with minimal use of purchased inputs. As is the case in other largely smallholder systems, livestock have a range of simultaneous roles in this system, including animal traction, production of manure, use as cash reserve, in addition to the production of meat and milk. Fuel-wood is often scarce as a result of deforestation and range degradation, leading to the ever-increasing role of animals as providers of manure for fuel, in addition to means of transport.

Issues and Perspectives

While this system supports larger populations than any grazing system, only 10 percent of the world population is related to this system. Fifty-one percent of the population involved is in Asia, mainly India, and 24 percent in the West Asia and North Africa region.

The major concern related to this system is the degradation of land resources, due to their limited production potential under growing population pressure. In livestock terms, this relates particularly to overgrazing and range degradation. This is connected to increasing stock numbers but also to crop production being expanded into increasingly marginal lands.

Given the extensive livestock rearing practised, livestock in the MRA system produce relatively high amounts of methane per animal kept and more so per kilogram of meat or milk produced.

There is a close interaction with the LGA system. With increasing population pressure, the LGA system tends to evolve into mixed systems, mainly the MRA system, due to the greater caloric efficiency of cropping *vis-à-vis* ruminant production when land becomes scarce. The outlook for this system is relatively similar to the one for the LGA system. The resource base puts a clear ceiling to agricultural intensification. Low and variable response to inputs makes their use financially risky. Population growth in this setting is leading to over-exploitation of the natural resource base, as traditional property rights cannot cope with the growing demands on the resource base. Alternative development strategies and the reduction of population pressure on the resource base are key elements for the sustainable development of these regions.

In the past, irrigation has been seen as the logical strategy to cope with the central constraint to agricultural production in this region, i.e. low and variable rainfall. Results have been mixed at best. Some reasons for the frequent failures were the high investment, the length of the training required to educate rainfed farmers in efficient irrigation management, the short useful life of many irrigation schemes due to salinization. Furthermore, best locations for irrigation schemes have already been exploited by now. Thus, a blend of other strategies is required in these regions, which involves promoting the mobility of workers to other regions and sectors, the *in situ* development of other sectors of the economy such as mining, tourism, fisheries, etc.

Mixed Irrigated Systems

The geographic distribution of the mixed irrigated system, the mixed irrigated system contributes about 23 percent of the total meat production worldwide.

Temperate Zones and Tropical Highlands (MIT)

This system belongs to the group of the land-based mixed systems of temperate and tropical highland regions. The peculiar feature is the existence of irrigation, which strongly influences the feed availability for ruminants and the variability of crop production. This changes the production environment substantially and determines the competitiveness of animal production *vis-à-vis* crop production in a given location. This system is found particularly in the Mediterranean region (Portugal, Italy, Greece, Albania, Bulgaria) and in the Far East (North and South Korea, Japan, and parts of China). These are agro-ecologies in the transition between subtropical and temperate conditions, where plant growth is limited, both by low temperatures in the cold season and by moisture availability during the vegetation period. Their importance in tropical highlands is negligible.

Typical cases are south European family farms combining one cycle of irrigated crop production with livestock production based on the grazing of drylands, crop stubbles and some irrigated alfalfa.

The transition to mixed irrigated arid systems is gradual, with the latter having year-round production on irrigated land, thus reducing the opportunities for grazing crop stubbles. Far-east Asian mixed family farms are mainly based on irrigated rice and dairy cattle.

Resources and Production

Traditional local sheep and cattle breeds have been largely displaced as management practices and product prices allowed for more intensive production and the associated increase in the use of external inputs (energy for water pumping, fertilizers, agrochemicals.) In the Mediterranean region, the main feed resource has traditionally been the silvopastoral system, supplemented by crop by-products. In the land-scarce, intensive East Asian systems, the main resources are cereal straw, intensively managed pastures, forages and imported feeds.

Livestock production technology is basically the same utilized by the MRT-system. High product prices and a high opportunity cost for labour make intensive production systems viable. This implies a heavy effort to actively adjust seasonal feed supply to the rather constant requirements of the herds and flocks. This is achieved through forage conservation (hay, silage) and through the feeding of grains and grain by-products.

In the more extensive situations, such as in the Chinese MIT system, the integration of livestock into the farming system is broader in physical terms. Animal traction is an important input into the crop system. Less productive animal breeds are fed less concentrate feeds and therefore consume more crop by-products. Manure is actively allocated to the more productive irrigated fields thus transferring nutrients from other parts of the farm to the irrigated fields.

Weeds are Fed to the Ruminants

Meat, milk and wool, the main outputs of this system, are mainly produced for the market. Manure is an issue only where animals are stabled, at least for certain periods of the day or the year. Animal traction has been displaced completely by engine-powered equipment in developed countries and the MIT system in China is gradually following the same path. Pigs, ducks, geese and chicken play a minor role, mainly in LDCs in utilizing crop by-products and family labour.

Issues and Perspectives

About 10 percent of the world population live in regions where this system is dominant. A large share of them belong to developed countries with relatively high income levels and where agricultural trade is important. This system tends to be found in regions with rather high population density. The major issue in environmental terms is the use of water, with agriculture competing with the use for urban supply. Another important issue is the management of the lands that are not irrigated. Particularly in the Mediterranean region, complex silvopastoral systems have been developed combining rainfed tree crops (olive trees, hazel nuts, cork-oaks) with extensive grazing, mainly of small ruminants.

Interactions with other systems are mainly trade related and are expected to increase in the future as agricultural protection is reduced. This competition will be mainly with mixed rainfed temperate systems, which produce largely the same commodities. This system is clearly associated with very intensive agriculture in temperate regions with a high population density. This is the case of the Far East and the southern European regions.

They are producing typical commodities of temperate environments at very high levels of intensity. It is related to the historical land scarcity and to policies heavily protecting domestic agriculture. With the outcome of the General Agreement on Tariffs and Trade (GATT) negotiations, it can be expected that these systems will be less and less viable, having to compete with very efficient rain-fed systems producing the same commodities. The system can be expected to shift to more extensive production, using less water and chemical inputs. This will reduce the negative impacts of the system on the environment. The expansion of international trade and particularly the incorporation of southern European countries into the EU, has led to an increase in the intensive production systems of off-season vegetables and fruits on the best irrigated land. The integration with livestock has been reduced, with ruminant grazing systems declining in absolute terms and concentrating on the marginal sites.

Humid and Sub-humid Tropics and Sub-tropics (MIH)

This is a mixed system in tropical and subtropical regions with growing seasons of more than 180 days, in which irrigation of crops is significant. The MIH system is particularly important in Asia. High population densities require intensive crop production and the irrigation of rice makes it possible to obtain more than two crops per year, even under conditions of very seasonal rainfall, substantially reducing yield variability as compared with the yield of upland rice or other rain-fed crops. Animal production has in the past been closely linked to the animal traction issue. In many Asian countries, smallscale mechanization is replacing it now, releasing feed resources for animal production to the markets. Typical cases are irrigated rice-buffalo systems of the Philippines, Vietnam, etc.

Resources and Production

Buffaloes and cattle have mainly been selected for animal traction in this system, involving both tillage and transportation. As mechanization expands, these animals selected mainly for the adaptation and animal traction performance may gradually be substituted by highly productive breeds to respond to a growing demand for meat and, to a lesser extent, dairy products. Pigs and poultry (particularly ducks and geese) play an important role in utilizing otherwise lost feed resources. Potentially valuable genes of adaptation to high fibre diets, tolerance to diseases, etc may be at risk in this system. Given the land scarcity, the major feed resources comprise crop by-products, straws, brans, weeds and roadside pastures. High yielding varieties of rice have emphasized grain production, frequently at the expense of their contribution to animal feed production (quality and quantity of straw, use of herbicides to control competition from weeds, etc). Highly productive forages for cut-and-carry systems, capable of growing on non-irrigated land are a potential avenue for intensification. Short term forage crops relay-planted into rice fields are also being tested. Tuber crops such as cassava and sweet potatoes, capable of producing acceptable yields of feeds of high energy concentration per kilogram of dry matter are an important resource for pig and to a lesser extent poultry production. The high productivity of land in this system is achieved through intensive land use of irrigated areas. Hence, the need for animal traction or mechanization to rapidly till the land after harvest to achieve a new crop cycle. This clearly limits grazing of stubbles and explains the efforts to harvest straw and treat it for feeding ruminants. Cattle and buffaloes are mainly tethered or fed cut-and-carry forages. Ducks are to some extent fed on insects in rice fields, a system in conflict with the increasing use of insecticides in rice production.

The main contribution of ruminants to this system has been animal traction. This function is gradually being taken over by small-scale machinery. Gradually, ruminants are assuming the role of providers of an additional cash income, a way to convert fibrous crop by-products and slack family labour into marketable livestock products, which are increasingly demanded by urban dwellers. Pigs and poultry provide meat for both home consumption and for the growing urban markets. MIH systems throughout the world produce 13 million tonnes of pork (18 percent of global production), more than any other land-based tropical system. Manure is recycled on the fields.

Issues and Perspectives

Among the tropical and subtropical systems, the MIH system is the one related to the largest population group, 990 million people, 97 percent of which are in Asia.

The environmental issues are related to the hygiene risks involved in keeping animals very close to people in areas of high population density. System-wide environmental issues are the frequently low efficiency of water utilization, and related erosion problems and the production of methane from paddy fields.

Competition for urban markets for livestock products is the main form of interaction with the landless monogastrics system, both domestically and globally through international trade.

This system has developed under high population pressure into a very closed system, capable of sustaining the basic needs of a large population. The challenge is how to maintain its sustainability in a changing setting: economic development is creating alternative employment and raising the opportunity cost of labour, consumers are purchasing increasing quantities of animal products and expecting products of different attributes: less fat, more homogeneous characteristics, more processing, etc. At the same time, expanding international trade is providing opportunities to access low cost feeds. These trends are promoting a certain degree of specialization while environmental concerns favour the maintenance of the traditional highly integrated system.

Arid and Semi-arid Tropics and Sub-tropics (MIA)

This is a mixed system of arid and semi-arid regions, in which irrigation makes year round intensive crop production feasible. It is found in the Near East, South Asia, North Africa, western United States and Mexico. Typical cases are luzerne/maize-based intensive dairy systems in California, Israel and Mexico; small-scale buffalo milk production in Pakistan; and animal traction based cash crop production in Egypt and Afghanistan.

Resources and Production

Cattle and buffaloes for milk and animal traction are the main ruminant resource. Sheep and goats are important where marginal rangelands are available in addition to irrigated land. In the MIA system, pigs are kept only in the Far East; they are virtually inexistent in West Asia and North Africa, largely for cultural reasons (Islamic and Jewish religions only). The main introduced breeds are dairy cattle to supply milk to large urban centres. Under good management conditions, intensive dairy schemes have been quite successful in hot but dry environments. Some of the world's highest lactation yields are achieved in the MIA system in Israel and California. The traditional smallholder MIA system in Asia relies heavily on buffaloes for milk production.

Luzerne is the forage crop favoured for use under irrigated conditions, due to the plant's capacity to colonize and improve desert soils brought into irrigation schemes. Furthermore, luzerne is high yielding and of high quality, a fact which makes it particularly suitable to supplement ruminant rations based on straws of low digestibility. Straw from irrigated crops is an important feed resource. In this system, efforts to treat straw to increase digestibility are quite attractive. Under developed country conditions, ample use of concentrates is made to feed high production dairy cows.

Milk production management in the MIA system is highly diverse, ranging from traditional buffalo management in backyards fed mainly cut-and-carry forages and straw to large-scale dairy farms milking several hundred cows, mainly Holstein Friesians. In this case, herd management is aided by computer programmes determining management interventions such as daily levels of concentrate supplementation, timing of drying, vaccinating, pregnancy checking, etc.

In the traditional MIA system, irrigated crop production is the main source of income with livestock playing a very secondary role. This is generally reflected in rather extensive management of the livestock enterprises.

Using irrigable land for forage production tends to be economical only for relatively efficient milk production, if an attractive urban market for fresh milk and dairy products exists. This is the case when imports of dried milk and dairy products are restricted or consumers are willing to pay a premium for products made from fresh milk *vis-à-vis* those based on reconstituted milk. Elsewhere, MIA systems are cash crop oriented and large ruminants are kept mainly for animal traction. Furthermore, fuelwood tends to be a scarce resource in these systems, a fact frequently leading to the use of manure as fuel.

Issues and Perspectives

The MIA system is predominant in regions that are home to over 750 million people, two thirds of them in Asia and one third in West Asia and North Africa. A large proportion of the total labour input into these systems is allocated to irrigated cash crop production. Milk production is mainly located in the proximity of urban centres. Particularly in modern, large scale operations, manure disposal tends to be an environmental problem. The main draw back of the MIA system, however, is water use, deficient drainage and salinization of irrigated land. The existence of certain fodder crops that tolerate relatively high levels of salinity, opens an avenue for livestock production as a strategy to live with the problem of salinization. The main interaction with other systems occurs through the international market, particularly for milk and dairy products. The MIA system makes an important contribution to food availability and employment in semi-arid and arid regions. The long term sustainability of these systems is nevertheless challenged by the problem of salinization of soils. Livestock play only an ancillary role, which may even decline in the development process, as appropriate mechanization becomes economically viable and as freer international trade and better infrastructure enhance the opportunities for consuming livestock products produced within more suitable environments.

Landless Systems (LL)

The developed countries dominate the picture of landless intensive production with more than half of total meat production. Asia is already contributing some 20 percent and eastern Europe 15 percent, with the latter recently in sharp decline.

Monogastric Production System (LLM)

Definition and Geographical Distribution

This system is defined by the use of monogastric species, mainly chicken and pigs in a production system where feed is introduced from outside the farm, thus separating decisions concerning feed use from those of feed production, and particularly of manure utilization on fields to produce feed and/or cash crops. Thus, this system is an open system in terms of nutrient flow.

The importation of nutrients normally occurs via markets, also international markets. While the return of nutrients through manure frequently causes problems given the high water content and thus high cost of transporting those nutrients to land-based systems capable of using them. On the other hand, mineral fertilizers are frequently a cheaper source of nutrients, thus reducing demand from other production systems for this resource, therefore turning it into "waste". Thus, the disposal of manure creates a major environmental impact of this system, particularly when production takes place close to highly populated urban centres. It also adds dimensions of pollution by odours and human health risks. Landless monogastrics systems are found predominantly in OECD member countries with 52 percent of the total landless pork production and 58 percent of the landless poultry production globally. In the case of pig production, Asia is second, with 31 percent of the world total. For poultry, Central and South America follow with 15 percent. To a large extent, this geographical distribution is determined by markets and consumption patterns, in addition to urbanization levels.

In developed countries with abundant road and cooling infrastructure, large-scale landless operations are located close to ports in net grain importing countries such as pig operations in the Netherlands or northern Germany. In grain exporting countries, such as the United States, landless systems tend to be located in grain producing areas, such as the states of Iowa, Illinois, etc. In countries with less developed infrastructures, such as roads and chilling, these operations are close to major urban centres, reflecting the feasibility of transporting grains *vis-à-vis* animal products.

Resources and Production

The landless monogastric system is almost exclusively based on hybrid and high producing, exotic breeds. This genetic material is widely traded internationally. The expansion of this system is clearly linked with the extinction of traditional breeds. The system is frequently stratified, implying that different enterprises specialize in the production of parent material, the production of young animals or the fattening process. The short production cycle of these species implies a high turnover and therefore a capacity to rapidly adjust to changes in demand for the products and to prices of inputs. It also implies that stock numbers are a poor indicator for the importance of the sector. The system is characterized by an ample use of feeds of high energy concentration (mainly cereals, oilseeds and their by-products). This feature is central to understanding the rapid growth of the system worldwide. The high energy concentration allows transport of feeds over longer distances. This provides for the expansion of production pulled by market incentives, based on imported feeds. Production of feeds is separated from their utilization. Transportation of concentrate feeds can be achieved at substantially lower costs than that of perishable animal products, even though the quantities are larger. Furthermore, consumers tend to pay a premium for fresh animal products *vis-à-vis* frozen/preserved products. Seasonality of feed production is easily overcome through grain storage and/or deferred purchasing on the market. The system is very knowledge- and capital-intensive, easily transferred across agroecological conditions, given the scarce links to the land base. Production efficiency is high in terms of output per unit of feed or per man-hour, less so when measured in terms of energy units. Concentrate conversion rates range between 2.5 - 4 kg/kg of pork, 2.0 to 2.5 kg feed DM/kg of poultry meat, and even lower for eggs. Capital intensity is high in all cases but wide variations are found. Very sophisticated automated systems are used in developed countries, responding to high labour costs. Variability of production within individual enterprises over time is low as long as management systems in place control exogenous factors correctly, i.e. disinfection, isolation from animals external to the system, effective quality control of feed inputs, etc.

Capacity of traditional breeds to cope with these challenges has been replaced by ability to perform at higher levels of efficiency in terms of desired outputs, as long as these external challenges are controlled by management. Management and infrastructure requirements generate large economies of scale in these systems. This implies large herd/flock sizes, large volumes of wastes and high animal health risks. Since products of this system are almost exclusively geared to urban markets, they have to comply to standardization and other specific quality criteria to be efficiently transported, processed and marketed. Many of these criteria are determined by the processing industries, rather than by the final consumers *per se.*

Issues and Perspectives

Given the tradeable nature of the inputs and the animal products involved in these systems, this system cannot be related to specific populations. Consumers of the system's outputs are mainly urban populations, frequently close to where the production base is located but also in other urban settings due to the active trade. The large-scale nature of the system and the heavy investments lead to very high labour productivity but very low employment. Thus, this system produces outputs for a large number of urban consumers but generate employment for few people. This employment tends to be relatively stable over time due to the low seasonality of production. While the employment effect at the production level is low, it must be acknowledged that the forward linkages in processing, wholesaling and retailing, as well as the backward linkages in inputs and services required, generate additional employment. The most important interactions with the environment are generation of large volumes of wastes and air pollution, as well as the increased demand for cereals, with the impact of the latter on the land resource base. In addition, the genetic erosion related to traditional breeds of chicken and pigs is of concern. Finally, given the character of substitutes of ruminant meats, it can be argued that the rapid development of "modern" landless monogastric systems has reduced the market incentives to expand ruminant production, thus reducing pressures for deforestation and degradation of rangelands. The system is typically competing with traditional land-based production for market shares in the urban markets. It must be kept in mind that poultry and pork are close substitutes for beef and mutton, thus also interacting with the ruminant systems. In a broader sense, the demand for cereals created by these systems is also competing for land resources with land-based ruminant systems.

Given the strong demand for these commodities, production can be expected to continue growing rapidly, particularly in LDCs. Landless poultry and pig production systems account for the majority of the output in developed countries and are rapidly increasing their share in LDCs given their high supply elasticity in the short run. The landless monogastrics system is an open system where important market failures imply a need for regulations. The negative impacts related to waste management are generally clearly located and regulations as well as technological innovations are mitigating the negative effects, particularly in developed countries. An important trend is the move to select more appropriate sites for production, away from urban centres to where enough land is available to make manure disposal through farming feasible.

The environmental impacts of these systems related to their high derived demand of cereals are of a global nature, given the links of these systems to the international grain markets.

Ruminant Production System (LLR)

Definition and Geographical Distribution

This system is defined by the use of ruminant species, principally cattle and marginally sheep, in production systems where feed is mainly introduced from outside the farm system, thus separating decisions of feed use from those of feed production and particularly of manure utilization on fields to produce feed and/or cash crops. Thus, this system is very open in terms of nutrient flow. It shares this feature with the landless monogastrics system. The main difference is that ruminants need more fibrous rations and that the feed conversion of concentrates to liveweight gains is substantially lower. These systems are only competitive under market conditions where consumers can afford to pay a substantial premium for quality beef over chicken or pork.

Information on ruminant meat production. Milk production has not been included in the quantitative analysis because the border between landless and landbased production is particularly blurred as roughage is essentially required to produce milk from healthy cows. In many cases, transport of roughage over a certain distance is economic. Thus the system description includes considerations on milk production but no quantitative estimates are provided.

Landless ruminant production systems are highly concentrated in a few regions of the world. In the case of cattle, they are almost exclusively found in eastern Europe and the CIS and in a few OECD member countries. Landless sheep production systems are only found in western Asia and Northern Africa.

Typical cases are large-scale feedlots in the United States and in eastern Europe and the CIS. Intensive dairy operations in the same regions are more land-based, due to the need to feed palatable fodder, which cannot be transported economically over long distances. Small-scale peri-urban dairy production, frequent in many LDCs, particularly in Asia, was not included in this system due to its very distinct nature, where manure is frequently recycled to home gardens or used as fuel and feeds which are mainly roughage produced close-by. These types of production are considered under mixed small-holder systems.

Examples are sheep fattening in Syria, feedlots in Texas and large-scale dairy operations in Eastern Europe.

Resources and Production

The landless ruminant system is based almost exclusively on high producing, specialized breeds and their crosses which, nevertheless, have not been bred specifically for performance under "landless" conditions. Furthermore, the limited proportion of total animals in these systems indicate that displacement of traditional breeds cannot be attributed specifically to this system. With regard to milk production, the Holstein Friesian breed is clearly the most important one as well as for beef production, English breeds predominate in the United States, while large European dual-purpose breeds provide animals for fattening. This clearly reflects the overall endowment with land and particularly range. The abundance of rangeland in the United States has led to the specialized production of calves from beef breeds for feedlot operations, while under European conditions these animals are a joint product with milk, mainly from mixed systems.

Apart from the high energy concentration feeds such as grains, this system requires fibrous feeds to maintain the rumen functions. This is frequently achieved through the use of silage, hay or fresh chopped forages. This requirement increases the complexity of these systems. To a large extent, ruminants are used like monogastric animals and their capability of efficiently utilizing fibrous feeds, not suitable for direct use by humans, is neglected. This is particularly true for the brief fattening process in North American feedlots, which improves carcass quality of young animals raised mainly on range of low opportunity cost.

This system is producing 12 percent of the global beef production. Production is highly concentrated in developed countries, mainly eastern Europe, the CIS and OECD member countries. The production system is highly capital-intensive, leading to substantial economies of scale. It is also feed-intensive and labour-extensive. Key efficiency parameters are daily weight gains and feed conversion, basically reflecting the efficiency in the use of capital invested in infrastructure or in the form of lean animals and feeds. Weight gains are usually in the range of 1 to 1.5 kg/day and feed conversion rates are about 8 to 10 kg of grains per kg of weight gain. In market oriented systems, such as the North American feedlot operations, economic performance is largely related to the evolution of prices of lean *versus* fat animals. Profitability is highest when the price differential for fat animals is large, as this effect is reflected in the price obtained for the total liveweight sold and not only for the additional weight gained in the feedlot. To avoid the downside risk involved in these price fluctuations, feedlot operators often hedge the risk through the option market. In this system products are almost exclusively geared to urban markets. In the case of the high quality beef produced, there is very limited processing involved. The situation of milk is more similar to the one of poultry and pork, a large and growing proportion being processed into dairy products.

Issues and Perspectives

Direct employment effects of this system are limited. Some additional employment is generated in specialized services and inputs required (particularly feed production, transportation, processing, supply of feeder cattle) as well as in the processing and marketing of the products.

These systems are competing for market shares and resources with all other livestock production systems through the cross-price elasticities for different meats and animal protein sources. Given the dependency on cooling and road infrastructure of trade in fresh animal products, this competition is stronger in developed countries than in countries with poor infrastructures. For the same reason, competition is stronger in cities close to ports than in the hinterland.

This system is closely linked to land-based systems that normally provide the young stock for landless systems. This constitutes an important difference to landless monogastrics systems, in which replacement stock is produced within the same system. These interactions are basically the same as those of landless monogastric systems. The most important ones are related to the production of animal wastes, leading to water and air pollution, acid rain and human health hazards. Furthermore, these systems induce extensive use of cereals with related environmental concerns (degradation of soils, nutrient transfers, use of agrochemicals, etc).

The landless ruminant system is producing only a small fraction of the ruminant meat output in developed countries and is of negligible importance in LDCs. It is critically dependent on high prices paid for quality beef and milk, and on ample supplies of low cost grains. The landless ruminant system can be expected to continue growing slowly in North America, driven by population growth, but with *per caput* consumption of beef stagnating. Its importance can be expected to decline in the European Community as production becomes more extensive in response to policies of reducing support to agriculture and promoting environmentally friendlier production systems.

The situation in eastern Europe and the former USSR is different. There, this system developed under central planning as an industrial process to produce these goods. With the shift to market economies, its importance is declining and ruminant production in that part of the world is shifting to the land base and to a smaller scale. A growing market for grain-fed beef exists in Japan and the newly industrialized countries of Asia. The growth rate of this market will depend mainly on the evolution of the international price of cereals and the increased *per caput* incomes. This market will in part be supplied domestically and through exports from the United States, Canada, Australia and possibly South America.

Globally, the system will continue to be of limited importance and mainly concentrated in the United States and in a few high income, arid countries of the Near East.

Issues and Trends

Only 9.3 percent of total meat is produced in grassland-based systems, compared to 36.8 percent in landless systems and 5.3 percent for mixed farming systems. One system alone, LLM accounts for more than half (52.3 percent) of global monogastric meat production.

Similarly, the MRT system alone produces more than half (55.5 percent) of total milk production from all considered species and 62.6 percent of cow milk production. Egg production is even more concentrated: more than two-thirds of total production is generated by one system, LLM. The landless intensive systems, particularly the one producing monogastric meat, are the fastest growing meat producing systems. Their growth rate of 4.3 percent p.a. compares with 2.2 for the mixed systems and 0.7 for the grassland based systems.

Compared to cropping systems, the system-to-system evolution of livestock systems alone is much less complex. Originally, all livestock production was basically grassland-based. Where climatic, soil and disease conditions permitted, grassland-based systems developed into mixed farming systems which covered a wide range of intensities and production modes as described earlier. The process was basically driven by population density as were the various forms of interaction between the crop and livestock sub-systems. Wherever urbanization and income exceeded certain levels, landless systems developed in the vicinity of urban centres, capitalizing on the efficiency and supply elasticity of these systems. Similar to crops, two primary sources of production growth can be distinguished: expansion in livestock numbers through an enlargement of the feed resource base, i.e. through an increased intensity of range and pasture utilization and higher use of feed concentrates and agricultural by-products; and higher output of meat, milk or eggs per animal through improved management, feeding, breeds and animal husbandry technologies. For pigs, poultry and to a lesser extent dairy cattle, much of the increased output comes from landless intensive or mixed farming systems and the use of concentrate. Successes in grain production, on the one hand, and the limited potential for cost effectively increasing quantity and quality of feed from extensive range on the other lead to shift of growth from range to grainbased, production systems, particularly in Asia where growth rates are highest. The slow evolution from extensive to intensive production has led to increases in environmental degradation, particularly for grazing systems and mixed farming with low degrees of integration. There are considerable institutional and economic problems to be overcome in bringing livestock numbers into balance with forage and feed availabilities. These problems will be difficult to overcome in the short to medium term and are likely to grow in scope and gravity. Livestock production takes place under very diverse conditions in the different developing countries. However, the direction of change, even if gradual, is towards more intensive production with less dependence on open range feeding which imposes excessive burdens on the environment, and with improved and balanced feeding practices and improved breeds. The improved practices enables more of the feed to go production rather to inefficient maintenance. This has led lead to progresses in feed conversion efficiency.

Intensification must outweigh expansion if livestock production is to respond to effective demand for livestock commodities. Livestock production systems differ in ability to respond to extra demand for livestock products which is primarily due to the biological characteristics of the production process. As an example, poultry meat production has been demonstrated to be quickest in responding to increasing demand. This is due to the industrial type of production in which this commodity is produced, increasingly also in the developing countries. Fast reproduction cycles allow to react to changing demand within months. A wide range of commodities can be diverted from other uses (food, industrial uses) to feed, if prices permit - this allows for a fast adaptation of the feed resource base for poultry as opposed to ruminants. The demand for land is low which allows the establishment of production units close to consumer centres. High feed conversion efficiencies make poultry production a highly profitable enterprise.

There are other production systems which are also, to a large extent, driven by consumer demand (eggs, then pork, and to a lesser extent dairy) and the size and type of production units for these commodities are largely driven by market forces. The rate of expansion and intensification of these production systems usually does not allow a gradual transformation from traditional production into these modern types. The growth process here tends to be not evolutionary but discontinued. For red meat, the pace of growth is such that in most cases the evolution of traditional production systems into more intensified modes of production is through adaptation. The type of commodities (beef, mutton and goat meat, and, to a lesser extent, also dairy) lends itself to a more gradual transformation because of the long reproduction cycle, low feed conversion efficiency, and a lower degree of specialization. This is mainly due to the physical characteristics of the production process.

Traditional, integrated mixed farming systems are also unlikely to be highly responsive to increasing demand simply because of the other functions that livestock have to fulfil within the farm-household system. The production reserves in the form of productivity increases and production enlargement that can be realized through a gradual transformation of traditional farming systems are usually insufficient to respond effectively to growing demand. As a result, almost all developing countries experienced the emergence of modern production systems, similar to those in the developed countries. As traditional systems prove increasingly unable to meet the rising demand, an increasing share of total supply will come from highly intensive systems.

Among the regions, growth rates have been highest in Asia with a 7.8 percent increase annually in beef and veal production, 6.3 percent in sheep and goat meat, 7.0 percent in cow milk production, 7.0 percent in pork, 9.6 in poultry meat and 9.6 percent in poultry eggs. Comparing these growth rates with stock increases where possible (for ruminants) it shows that there is very little horizontal expansion (1.2 percent for cattle stock, 1.8 percent for sheep and goats, 2.3 percent for cattle dairy stocks). This means that annual productivity increases of between 4 and 6 percent have been obtained in the decade. This is unprecedented and can only be compared to the Green Revolution in crop production in the same region during the 1960s and early 1970s. For ruminants, the growth has taken place almost exclusively in mixed farming systems and landless ruminant systems have not developed to any significant scale. On the contrary, landless monogastric systems are now found in every country of the region and are quickly developing in the vicinity of urban centres. This, in analogy to the growth path precedented by OECD member countries, is driven by increasing *per caput* incomes, urbanization and changes in consumption patterns. Unless full sets of regulatory measures are established and enforced peri-urban landless systems will continue to grow quickly, responding to the surging demand of low-priced livestock products.

Chapter 2

Livestock and Sustainable Agriculture

Agriculture has changed dramatically, especially since the end of World War II. Food and fibre productivity soared due to new technologies, mechanization, increased chemical use, specialization and government policies that favoured maximizing production. These changes allowed fewer farmers with reduced labour demands to produce the majority of the food and fibre in the U.S. Although these changes have had many positive effects and reduced many risks in farming, there have also been significant costs. Prominent among these are topsoil depletion, ground-water contamination, the decline of family farms, continued neglect of the living and working conditions for farm labourers, increasing costs of production, and the disintegration of economic and social conditions in rural communities.

A growing movement has emerged during the past two decades to question the role of the agricultural establishment in promoting practices that contribute to these social problems. Today this movement for sustainable agriculture is garnering increasing support and acceptance within mainstream agriculture. Not only does sustainable agriculture address many environmental and social concerns, but it offers innovative and economically viable opportunities for growers, labourers, consumers, policymakers and many others in the entire food system. This chapter is an effort to identify the ideas, practices and policies that constitute our concept of sustainable agriculture. We do so for two reasons: 1) to clarify the research agenda and priorities of our program, and 2) to suggest to others practical steps that may be appropriate for them in moving toward sustainable agriculture. Because the concept of sustainable agriculture is still evolving, we intend the paper not as a definitive or final statement, but as an invitation to continue the dialogue.

Concept Themes

Sustainable agriculture integrates three main goals—environmental health, economic profitability, and social and economic equity. A variety of philosophies, policies and practices have contributed to these goals. People in many different capacities, from farmers to consumers, have shared this vision and contributed to it. Despite the diversity of people and perspectives, the following themes commonly weave through definitions of sustainable agriculture.

Sustainability rests on the principle that we must meet the needs of the present without compromising the ability of future generations to meet their own needs. Therefore, *stewardship of both natural and human resources* is of prime importance. Stewardship of human resources includes consideration of social responsibilities such as working and living conditions of labourers, the needs of rural communities, and consumer health and safety both in the present and the future. Stewardship of land and natural resources involves maintaining or enhancing this vital resource base for the long term.

A *systems perspective* is essential to understanding sustainability. The system is envisioned in its broadest sense, from the individual farm, to the local ecosystem, *and* to communities affected by this farming system both locally and globally. An emphasis on the system allows a larger and more thorough view of the consequences of farming practices on both human communities and the environment. A systems approach gives us the tools to explore the interconnections between farming and other aspects of our environment.

A systems approach also implies *interdisciplinary efforts in research and education*. This requires not only the input of researchers from various disciplines, but also farmers, farmworkers, consumers, policymakers and others.

Making the transition to sustainable agriculture is a process. For farmers, the transition to sustainable agriculture normally requires a series of small, realistic steps. Family economics and personal goals influence how fast or how far participants can go in the transition. It is important to realize that each small decision can make a difference and contribute to advancing the entire system further on the "sustainable agriculture continuum." The key to moving forward is the will to take the next step.

Finally, it is important to point out that *reaching toward the goal of sustainable agriculture is the responsibility of all participants in the system*, including farmers, labourers, policymakers, researchers, retailers, and consumers. Each group has its own part to play, its own unique contribution to make to strengthen the sustainable agriculture community.

The remainder of this document considers specific strategies for realizing these broad themes or goals. The strategies are grouped according to three separate though related areas of concern: Farming and Natural Resources, Plant and Animal Production Practices, and the Economic, Social and Political Context. They represent a range of potential ideas for individuals committed to interpreting the vision of sustainable agriculture within their own circumstances.

Farming and Natural Resources

Water : When the production of food and fibre degrades the natural resource base, the ability of future generations to produce and flourish decreases. The decline of ancient civilizations in Mesopotamia, the Mediterranean region, Pre-Columbian southwest U.S. and Central America is believed to have been strongly influenced by natural resource degradation from non-sustainable farming and forestry practices. Water is the principal resource that has helped agriculture and society to prosper, and it has been a major limiting factor when mismanaged.

Water supply and use: In California, an extensive water storage and transfer system has been established which has allowed crop production to expand to very arid regions. In drought years, limited surface water supplies have prompted overdraft of ground-water and consequent intrusion of salt water, or permanent collapse of aquifers. Periodic droughts, some lasting up to 50 years, have occurred in California. Several steps should be taken to develop drought-resistant farming systems even in "normal" years, including both policy and management actions: 1) improving water conservation and storage measures, 2) providing incentives for selection of drought-tolerant crop species, 3) using reduced-volume irrigation systems, 4) managing crops to reduce water loss, or 5) not planting at all.

Water quality: The most important issues related to water quality involve salinization and contamination of ground and surface waters by pesticides, nitrates and selenium. Salinity has become a problem wherever water of even relatively low salt content is used on shallow soils in arid regions and/or where the water table is near the root zone of crops. Tile drainage can remove the water and salts, but the disposal of the salts and other contaminants may negatively affect the environment depending upon where they are deposited. Temporary solutions include the use of salt-tolerant crops, low-volume irrigation, and various management techniques to minimize the effects of salts on crops. In the long-term, some farmland may need to be removed from production or converted to other uses. Other uses include conversion of row crop land to production of drought-tolerant forages, the restoration of wildlife habitat or the use of agroforestry to minimize the impacts of salinity and high water tables. Pesticide and nitrate contamination of water can be reduced using many of the practices discussed later in the Plant Production Practices and Animal Production Practices sections.

Wildlife: Another way in which agriculture affects water resources is through the destruction of riparian habitats within watersheds. The conversion of wild habitat to agricultural land reduces fish and wildlife through erosion and sedimentation, the effects of pesticides, removal of riparian plants, and the diversion of water. The plant diversity in and around both riparian and agricultural areas should be maintained in order to support a diversity of wildlife. This diversity will enhance natural ecosystems and could aid in agricultural pest management.

Energy: Modern agriculture is heavily dependent on non-renewable energy sources, especially petroleum. The continued use of these energy sources cannot be sustained indefinitely, yet to abruptly abandon our reliance on them would be economically catastrophic. However, a sudden cutoff in energy supply would be equally disruptive. In sustainable agricultural systems, there is reduced reliance on non-renewable energy sources and a substitution of renewable sources or labour to the extent that is economically feasible.

Air: Many agricultural activities affect air quality. These include smoke from agricultural burning; dust from tillage, traffic and harvest; pesticide drift from spraying; and nitrous oxide emissions from the use of nitrogen fertilizer. Options to improve air quality include incorporating crop residue into the soil, using appropriate levels of tillage, and planting wind breaks, cover crops or strips of native perennial grasses to reduce dust.

Soil: Soil erosion continues to be a serious threat to our continued ability to produce adequate food. Numerous practices have been developed to keep soil in place, which include reducing or eliminating tillage, managing irrigation to reduce runoff, and keeping the soil covered with plants or mulch. Enhancement of soil quality is discussed in the next section.

Plant Production Practices

Sustainable production practices involve a variety of approaches. Specific strategies must take into account topography, soil characteristics, climate, pests, local availability of inputs and the individual grower's goals. Despite the site-specific and individual nature of sustainable agriculture, several general principles can be applied to help growers select appropriate management practices:

- Selection of species and varieties that are well suited to the site and to conditions on the farm;
- Diversification of crops (including livestock) and cultural practices to enhance the biological and economic stability of the farm;
- Management of the soil to enhance and protect soil quality;
- Efficient and humane use of inputs; and
- Consideration of farmers' goals and lifestyle choices.

Selection of site, species and variety: Preventive strategies, adopted early, can reduce inputs and help establish a sustainable production system. When possible, pest-resistant crops should be selected which are tolerant of existing soil or site conditions. When site selection is an option, factors such as soil type and depth, previous crop history, and location (e.g. climate, topography) should be taken into account before planting.

Diversity: Diversified farms are usually more economically and ecologically resilient. While monoculture farming has advantages in terms of efficiency and ease of management, the loss of the crop in any one year could put a farm out of business and/or seriously disrupt the stability of a community dependent on that crop. By growing a variety of crops, farmers spread economic risk and are less susceptible to the radical price fluctuations associated with changes in supply and demand.

Properly managed, diversity can also buffer a farm in a biological sense. For example, in annual cropping systems, crop rotation can be used to suppress weeds, pathogens and insect pests. Also, cover crops can have stabilizing effects on the agroecosystem by holding soil and nutrients in place, conserving soil moisture with mowed or standing dead mulches, and by increasing the water infiltration rate and soil water holding capacity. Cover crops in orchards and vineyards can buffer the system against pest infestations by increasing beneficial arthropod populations and can therefore reduce the need for chemical inputs. Using a variety of cover crops is also important in order to protect against the failure of a particular species to grow and to attract and sustain a wide range of beneficial arthropods.

Optimum diversity may be obtained by integrating both crops and livestock in the same farming operation. This was the common practice for centuries until the mid-1900s when technology, government policy and economics compelled farms to become more specialized. Mixed crop and livestock operations have several advantages. First, growing row crops only on more level land and pasture or forages on steeper slopes will reduce soil erosion. Second, pasture and forage crops in rotation enhance soil quality and reduce erosion; livestock manure, in turn, contributes to soil fertility. Third, livestock can buffer the negative impacts of low rainfall periods by consuming crop residue that in "plant only" systems would have been considered crop failures. Finally, feeding and marketing are flexible in animal production systems. This can help cushion farmers against trade and price fluctuations and, in conjunction with cropping operations, make more efficient use of farm labour.

Soil management: A common philosophy among sustainable agriculture practitioners is that a "healthy" soil is a key component of sustainability; that is, a healthy soil will produce healthy crop plants that have optimum vigor and are less susceptible to pests. While many crops have key pests that attack even the healthiest of plants, proper soil, water and nutrient management can help prevent some pest problems brought on by crop stress or nutrient imbalance. Furthermore, crop management systems that impair soil quality often result in greater inputs of water, nutrients, pesticides, and/or energy for tillage to maintain yields.

In sustainable systems, the soil is viewed as a fragile and living medium that must be protected and nurtured to ensure its long-term productivity and stability. Methods to protect and enhance the productivity of the soil include using cover crops, compost and/or manures, reducing tillage, avoiding traffic on wet soils, and maintaining soil cover with plants and/or mulches. Conditions in most California soils (warm, irrigated, and tilled) do not favour the buildup of organic matter. Regular additions of organic matter or the use of cover crops can increase soil aggregate stability, soil tilth, and diversity of soil microbial life.

Efficient use of inputs: Many inputs and practices used by conventional farmers are also used in sustainable agriculture. Sustainable farmers, however, maximize reliance on natural, renewable, and on-farm inputs. Equally important are the environmental, social, and economic impacts of a particular strategy. Converting to sustainable practices does not mean simple input substitution. Frequently, it substitutes enhanced management and scientific knowledge for conventional inputs, especially chemical inputs that harm the environment on farms and in rural communities. The goal is to develop efficient, biological systems which do not need high levels of material inputs. Growers frequently ask if synthetic chemicals are appropriate in a sustainable farming system. Sustainable approaches are those that are the least toxic and least energy intensive, and yet maintain productivity and profitability. Preventive strategies and other alternatives should be employed before using chemical inputs from any source. However, there may be situations where the use of synthetic chemicals would be more "sustainable" than a strictly nonchemical approach or an approach using toxic "organic" chemicals. For example, one grape grower switched from tillage to a few applications of a broad spectrum contact herbicide in the vine row. This approach may use less energy and may compact the soil less than numerous passes with a cultivator or mower.

Consideration of farmer goals and lifestyle choices. Management decisions should reflect not only environmental and broad social considerations, but also individual goals and lifestyle choices. For example, adoption of some technologies or practices that promise profitability may also require such intensive management that one's lifestyle actually deteriorates. Management decisions that promote sustainability, nourish the environment, the community *and* the individual.

Animal Production Practices

In the early part of this century, most farms integrated both crop and livestock operations. Indeed, the two were highly complementary both biologically and economically. The current picture has changed quite drastically since then. Crop and animal producers now are still dependent on one another to some degree, but the integration now most commonly takes place at a higher level—*between* farmers, through intermediaries, rather than *within* the farm itself. This is the result of a trend toward separation and specialization of crop and animal production systems. Despite this trend, there are still many farmers, particularly in the Midwest and Northeastern U.S. that integrate crop and animal systems—either on dairy farms, or with range cattle, sheep or hog operations.

Even with the growing specialization of livestock and crop producers, many of the principles outlined in the crop production section apply to both groups. The actual management practices will, of course, be quite different. Some of the specific points that livestock producers need to address are listed below.

Management Planning: Including livestock in the farming system increases the complexity of biological and economic relationships. The mobility of the stock, daily feeding, health concerns, breeding operations, seasonal feed and forage sources, and complex marketing are sources of this complexity. Therefore, a successful ranch plan should include enterprise calendars of operations, stock flows, forage flows, labour needs, herd production records and land use plans to give the manager control and a means of monitoring progress toward goals.

Animal Selection: The animal enterprise must be appropriate for the farm or ranch resources. Farm capabilities and constraints such as feed and forage sources, landscape, climate and skill of the manager must be considered in selecting which animals to produce. For example, ruminant animals can be raised on a variety of feed sources including range and pasture, cultivated forage, cover crops, shrubs, weeds, and crop residues. There is a wide range of breeds available in each of the major ruminant species, i.e., cattle, sheep and goats. Hardier breeds that, in general, have lower growth and milk production potential, are better adapted to less favourable environments with sparse or highly seasonal forage growth.

Animal nutrition: Feed costs are the largest single variable cost in any livestock operation. While most of the feed may come from other enterprises on the ranch, some purchased feed is usually imported from off the farm. Feed costs can be kept to a minimum by monitoring animal condition and performance and understanding seasonal variations in feed and forage quality on the farm. Determining the optimal use of farm-generated by-products is an important challenge of diversified farming.

Reproduction: Use of quality germplasm to improve herd performance is another key to sustainability. In combination with good genetic stock, adapting the reproduction season to fit the climate and sources of feed and forage reduce health problems and feed costs.

Herd Health: Animal health greatly influences reproductive success and weight gains, two key aspects of successful livestock production. Unhealthy stock waste feed and require additional labour. A herd health program is critical to sustainable livestock production.

Grazing Management: Most adverse environmental impacts associated with grazing can be prevented or mitigated with proper grazing management. First, the number of stock per unit area (stocking rate) must be correct for the landscape and the forage sources. There will need to be compromises between the convenience of tilling large, unfenced fields and the fencing needs of livestock operations. Use of modern, temporary fencing may provide one practical solution to this dilemma. Second, the long term carrying capacity and the stocking rate must take into account short and long-term droughts. Especially in Mediterranean climates such as in California, properly managed grazing significantly reduces fire hazards by reducing fuel build-up in grasslands and brushlands. Finally, the manager must achieve sufficient control to reduce overuse in some areas while other areas go unused. Prolonged concentration of stock that results in permanent loss of vegetative cover on uplands or in riparian zones should be avoided. However, small scale loss of vegetative cover around water or feed troughs may be tolerated if surrounding vegetative cover is adequate.

Confined Livestock Production: Animal health and waste management are key issues in confined livestock operations. The moral and ethical debate taking place today regarding animal welfare is particularly intense for confined livestock production systems. The issues raised in this debate need to be addressed.

Confinement livestock production is increasingly a source of surface and ground water pollutants, particularly where there are large numbers of animals per unit area. Expensive waste management facilities are now a necessary cost of confined production systems. Waste is a problem of almost all operations and must be managed with respect to both the environment and the quality of life in nearby communities. Livestock production systems that disperse stock in pastures so the wastes are not concentrated and do not overwhelm natural nutrient cycling processes have become a subject of renewed interest.

The Economic, Social and Political Context

In addition to strategies for preserving natural resources and changing production practices, sustainable agriculture requires a commitment to changing public policies, economic institutions, and social values. Strategies for change must take into account the complex, reciprocal and ever-changing relationship between agricultural production and the broader society.

The "food system" extends far beyond the farm and involves the interaction of individuals and institutions with contrasting and often competing goals including farmers, researchers, input suppliers, farmworkers, unions, farm advisors, processors, retailers, consumers, and policymakers. Relationships among these actors shift over time as new technologies spawn economic, social and political changes.

A wide diversity of strategies and approaches are necessary to create a more sustainable food system. These will range from specific and concentrated efforts to alter specific policies or practices, to the longer-term tasks of reforming key institutions, rethinking economic priorities, and challenging widely-held social values. Areas of concern where change is most needed include the following:

Food and agricultural policy: Existing federal, state and local government policies often impede the goals of sustainable agriculture. New policies are needed to simultaneously promote environmental health, economic profitability, and social and economic equity. For example, commodity and price support programs could be restructured to allow farmers to realize the full benefits of the productivity gains made possible through alternative practices. Tax and credit policies could be modified to encourage a diverse and decentralized system of family farms rather than corporate concentration and absentee ownership. Government and land grant university research policies could be modified to emphasize the development of sustainable alternatives. Marketing orders and cosmetic standards could be amended to encourage reduced pesticide use. Coalitions must be created to address these policy concerns at the local, regional, and national level.

Land use: Conversion of agricultural land to urban uses is a particular concern in California, as rapid growth and escalating land values threaten farming on prime soils. Existing farmland conversion patterns often discourage farmers from adopting sustainable practices and a long-term perspective on the value of land. At the same time, the close proximity of newly developed residential areas to farms is increasing the public demand for environmentally safe farming practices. Comprehensive new policies to protect prime soils and regulate development are needed, particularly in California's Central Valley. By helping farmers to adopt practices that reduce chemical use and conserve scarce resources, sustainable agriculture research and education can play a key role in building public support for agricultural land preservation. Educating land use planners and decision-makers about sustainable agriculture is an important priority.

Labour: In California, the conditions of agricultural labour are generally far below accepted social standards and legal protections in other forms of employment. Policies and programs are needed to address this problem, working toward socially just and safe employment that provides adequate wages, working conditions, health benefits, and chances for economic stability. The needs of migrant labour for year-around employment and adequate housing are a particularly crucial problem needing immediate attention. To be more sustainable over the long-term, labour must be acknowledged and supported by government policies, recognized as important constituents of land grant universities, and carefully considered when assessing the impacts of new technologies and practices.

Rural Community Development: Rural communities in California are currently characterized by economic and environmental deterioration. Many are among the poorest locations in the nation. The reasons for the decline are complex, but changes in farm structure have played a significant role. Sustainable agriculture presents an opportunity to rethink the importance of family farms and rural communities. Economic development policies are needed that encourage more diversified agricultural production on family farms as a foundation for healthy economies in rural communities. In combination with other strategies, sustainable agriculture practices and policies can help foster community institutions that meet employment, educational, health, cultural and spiritual needs.

Consumers and the Food System: Consumers can play a critical role in creating a sustainable food system. Through their purchases, they send strong messages to producers, retailers and others in the system about what they think is important. Food cost and nutritional quality have always influenced consumer choices. The challenge now is to find strategies that broaden consumer perspectives, so that environmental quality, resource use, and social equity issues are also considered in shopping decisions. At the same time, new policies and institutions must be created to enable producers using sustainable practices to market their goods to a wider public. Coalitions organized around improving the food system are one specific method of creating a dialogue among consumers, retailers, producers and others. These coalitions or other public forums can be important vehicles for clarifying issues, suggesting new policies, increasing mutual trust, and encouraging a long-term view of food production, distribution and consumption.

Livestock Economics

Agricultural economics is the study of applying economic management principles to food farming. The result, ideally, is an agriculture industry that better understands efficiency, sustenance and market demand. The field of agricultural economics looks at all elements of food production and applies rational thought and planning as a whole. From crops, livestock, land usage and soil content, all aspects of farm life are examined, including how its connection to one another can be strengthened. Many times, this involves learning about the latest technology to help crops or livestock, but it also might require a knowledge of what has and has not worked in the past.

Agricultural economics is a relatively new field, considering the countless years that people have been farming. Interest began to mount in the early 1900s, when many economic thinkers around the globe began focusing attention to agriculture. Noting that the act of planting, harvesting and distributing crops and livestock was inefficiently performed, academics believed that farms around the world could greater yields and profits with a change of thought. Additionally, many universities and colleges opened agricultural economics programs with the intent of preparing students for a career in this field.

Careers in economical agriculture are as wide ranging as the crops produced around the world. The principles of farm economics, agricultural production and management can be directly applied to being a successful farmer, but there are a multitude of other options. Seed and chemical companies utilize agricultural economic thought in their production and development, grain elevator companies and equipment manufacturers must understand the economic landscape for each crop in order to stay relevant, and salesmen use agricultural economics to better serve their clients with the products they demand.

Since its inception, agricultural economics has helped further the science behind farming, too. Advances in food preservation and shipping techniques have allowed a myriad of fruits, vegetables and meats to reach grocery stores. Currently, many economists the implementation of microcomputers in agriculture as another step toward streamlining farmland economics.

Agricultural economics is not a term that fits neatly within a single definition. It is the accumulation of many schools of thought on practically every aspect of agriculture, from planting a seed to serving food on a dinner table. It contains many different careers and needs that are constantly evolving as technology and economic thought grow.

India has 187.38 million cattle, which is about 15% of the world cattle population, of which around 12.07% are crossbred. Tamilnadu, Maharashtra, Kerala, Uttar Pradesh, Karnataka and Punjab account for 60 percent of the crossbred cattle population. 96.62 million Buffalo population in the country comes around 56 percent of the world Buffalo population. Recently the Government has identified livestock sector as one of the sustainable areas in Agricultural sector. More than 75 percent of the rural households in India depend on livestock sector for supplementary income. This sector contributes 30 percent of the total income from agriculture in the country. In southern states it is more than 40 percent. Annual growth rate in livestock sector is more than four times in agriculture. Commodity share is 3.53 as against 3.5 in agriculture.

Analysis of the trends over the last two decades indicates that the growth in poultry and dairy sectors has exceeded the growth in cereal production. There are certain indications, which suggest that the demand led livestock growth is expected to continue and by 2020 more than 60 percent of meat and 50 percent of milk will be produced in the developing countries. China and India are likely to emerge as the primary producers of meat and milk respectively.

Even though India is the largest milk producing country yet productivity per animal is only less than 50% of the world average. This is mainly due to poor level of nutrition and low genetic potential for milk production and health care. The gap between demand and supply of green and dry fodder presents a challenge for fodder production in the coming years. Studies done by NDDB revealed that 45% of the milk is consumed as liquid milk, 28% as Ghee, 6% as Butter, Khoa and 7 5% as Dahi and 2.6% as milk powder. In general consumption expenditure on milk and milk products is next to cereals and is rising steadily over the years. Demand elasticity estimates of Indian dairy industry for milk and milk products combined are 1.65 in rural and 1.15 in urban India. (Dutta and Ganguly 2000). Moreover expenditure elasticity of demand for milk and milk products for lower income class is considerably greater.

Livestock markets in the world are changing dramatically. There are increasing concerns over the systems of livestock production and safety of livestock products. The picture in the developing country appears markedly different. Demand for livestock products is predicted to increase by 5% or more per annum. It can be met from different sources, largely domestic but also international sources. It is expected that the most of the projected increase in the livestock demand will be met from within the developing world. It will however require enormous increases in supply of feeds and other inputs, scaling up of livestock production, processing, distribution and marketing and much improved systems for assuring the quality and safety of livestock products for the consumers.

How poor will benefit from the increasing consumption of livestock products will depend on a number of policy, technology and research choices. There is evidence from several countries that even small holders are very competitive under the right circumstances (Delgado et al 2001).

Market opportunities for the poor can be greatly improved by paying specific attention to social equity and environmental issues. Selective investments in infrastructure, co-operatives, contract farming arrangements and other pro poor market mechanisms should be done. Public support is required for appropriate health and food safety systems that could benefit the poor. These changes will require new ways of doing business and strategic research that targets the priority concerns of the poor. There is great potential for increasing the assets, incomes and food safety of the poor from livestock production, marketing and consumption.

While giving emphasis to production, marketing of the livestock products are not given due attention. Scientific breeding, feeding, management and disease control have been given due importance but when the issue of sustainability arises marketing comes in to picture. Livestock products except around 15% of the milk produced are mainly traded through unorganized sector. In order to explore the rural market for livestock products our production strategy need to be market oriented. It should be in tune with local, domestic and international market. Production and branding of traditional livestock products, which fetches good price, should be promoted.

There are certain pre requisites for popularizing market oriented production strategy for livestock products. Cost of production has to be reduced considerably without affecting the quality of products to make it economically viable. Changes in the extension approach, market forecasting system, value addition, awareness on diseases affecting trade of livestock products, consumer behaviour, production of livestock products based on the demographic characteristics of the population, good manufacturing and retail practices, best production practices and implementation of food safety norms, etc need more emphasis.

Extension approach should be need based and participatory in nature. It should be an integrated systems approach having problem-solving dimension. This approach should make include all efforts to improve the production while giving due attention to marketing. Farmers need to be given more awareness on value addition, food safety norms, branding and problems and prospects of marketing including the four cardinals of marketing like product, price, promotion and place. As told by the world marketing Guru Philip Kotler in today's world of IT explosion branding is becoming more important than before. One needs to redefine the role of marketing as creating, communicating and delivering value to the consumer. Value addition of livestock products should be based on consumer needs and taste, like fat free milk for cardiac patients, chocolates for children, quality cheese for international market, etc.

Measures to Make Livestock Products Internationally Viable and Competitive

If the farmer has to sustain and if his products have to find a good market the cost of production has to come down and the quality of the products has to be improved. For that cost-effective farming system has to be adopted and efficiency of production has to go up. Veterinarians have a significant role to play. The new innovations in technology and marketing should be brought to the farmer's doorsteps.

Techniques that will reduce cost of production and enhance production per animal should be developed and should be viable and sustainable.

Genetic improvement must be achieved at the same time maintaining the genetic diversity. Selection of breeds adaptable to the existing agro-climatic conditions that can thrive on unconventional feeds and fodder should be given priority. Biotechnology tools should be effectively utilized to harvest maximum output with less extensive inputs. Conservation of locally available germplasm that has lot of good traits should be attempted. The production system that would optimise farmers' return through judicial use of farm wastes should be developed. Integrated farming with pig, duck, and fish. The enhancement of production should be centered on small farmers. Reducing the calving interval and a targeted minimum calving interval should ensure maximum return to the farmer.

Effective utilization of feed and crop residue and other agricultural products should be given paramount importance. Improving the availability of nutrients in feed for the animal using enzymes, growth promoters, use of chelated minerals etc. should be thought of. Feed is an important constraint for livestock production in India and Green fodder and paddy straw are scarce commodities. Effective planning should be made for collection, processing, storage and distribution of paddy straw at the same time enriching its quality. Seasonal surpluses should be exploited and effort should be made for uniform availability throughout the country. Concerted effort should be made to make use of the available land for fodder production. Effective utilization of available crop residues, industrial wastes and unconventional fodder should be made. There is good scope to develop meat industry and its export potential is very good. Livestock for meat should be identified, male animals should be specially reared for meat purpose and veal production should be augmented.

Utmost care should be taken to ensure the quality of the products. We cannot think of export of the dairy or meat products without assuring quality. Strict hygienic measures at the site of production, during transport and storage should be maintained. Quality assurance laboratories should be established for meat, milk, feed and other products. We should have real surpluses of livestock products. Value added products should be developed to capture the market.

Marketing

Marketing is an important aspect of any livestock system. It provides the mechanism whereby producers exchange their livestock and livestock products for cash. The cash is used for acquiring goods and services which they do not produce themselves, in order to satisfy a variety of needs ranging from food items, clothing, medication and schooling to the purchase of breeding stock and other production inputs and supplies.

A major objective of pastoral systems research (PSR) is to increase productivity and improve the standard of living of pastoralists. Interventions for increasing productivity generated by PSR will have costs and returns associated with their adoption. It is therefore essential that such interventions are not only technically feasible and socially acceptable, but also economically feasible. In other words, the incremental returns of the interventions must out-weigh the additional costs incurred in adopting them. Both input and output prices fluctuate over time. Researchers must, therefore, establish the sensitivity of interventions by establishing within what range of input and output prices they are stable. Time series data on prices pastoralists are paid for their livestock and livestock products as well as prices they pay for inputs are essential for such analysis. Unfortunately, in most African countries while time series data on input prices may be available' they are almost non-existent for livestock prices.

The collection-of time series data on product and input prices, however, is not a function of PSR scientists. This should be the responsibility of the ministries servicing the livestock and agricultural sector. The purpose of this paper is to underscore the necessity of studying livestock markets and routinely collecting time series data on: prices of livestock at local, regional and terminal markets. Based upon the experience of ILCA economists and other researchers in Africa it outlines a methodology, which can be easily adopted by the relevant ministries of African countries in setting up a systematic collection and analysis of livestock market information. The second section provides the rationale for undertaking livestock marketing studies. A simplified livestock marketing system model is presented in the third section. A methodology for regularly collecting time series data and for conducting in-depth studies on livestock marketing is then given, followed by illustrated suggestions for the analysis of data. A suggested national organizational framework for collecting end analysing livestock market data is presented in the last section.

To the astute livestock producer/marketer, marketing means more than just selling. Astute marketing involves the entire planning process required to produce, promote or merchandise and price a commodity.

The first step in the marketing process is producing the type of stock - hogs, beef cattle, or sheep - that the livestock producer wants to produce. Of course, thought, the producer must produce the kind of animals that the market place wants. It is often a costly lesson to produce an animal that the market place doesn't particularly want.

Other essential components of the production and marketing process include: estimating production costs, calculating cash flow needs, knowing what type and quality of animal has been produced and which buyers will be interested in that type of animal. A final, and equally important, step in the plan is evaluating the pricing and delivery alternatives. Once the final sale has been made, it's very important to review the marketing process to determine what worked well and what needs to be improved. This module outlines how the marketing process can be organized into a number of very logical steps. The "Livestock Marketing" section of the marketing manual contains relevant modules that focus on these individual marketing steps in more detail.

Steps to Livestock Marketing Success

Estimate Costs

The first step involves accurately estimating costs of production and cash flow needs. This step is listed first because it is vitally import. Even though figuring costs and cash flow needs can be done at any time, it really is best to complete this step as early as possible. By estimating both production costs and cash flow requirements, a producer can decide what type of animal to produce and when it will have to be sold to meet payment schedules. These estimates, along with price forecasts, should be used to determine how the animal will be marketed. A producer who knows his or her past production costs and future price forecasts, can also determine when to retain female stock for breeding expansion or when to cull more heavily. Breakevens or production cost estimates are critical in setting a series of target prices that should be watched for in the changing market.

Gather Market Information and Including Market Outlook

Following market trends and projected livestock prices helps a livestock producer decide what to produce in order to bring the greatest return. For example, deciding whether to sell weaned calves, yearlings or slaughter cattle depends upon the market outlook for each of these animals. A producer may follow the United States market for price signals that may relate to our market. By following the US market, a producer may detect a market trend or even identify an export opportunity to a US market. For example, a decision could be made to finish animals to specifications required by an American buyer.

Know Your Product

The quality and type of livestock for sale must be assessed before a producer can seriously evaluate the various pricing and delivery alternatives. A producer that knows exactly what kind and quality of animal he or she has for sale, can contact buyers with the information they need. If there are premiums offered in the market for your type of cattle or other livestock, you will be better able to capitalize on them.

Knowing your product also involves presenting them favourably. Sorting animals into lots of similar size and weight will make them more attractive to buyers. Selling clean and healthy animals helps in reassuring buyers they are paying for a quality product.

Set Several Target Prices

Setting target prices is a big help in making livestock marketing decisions. However, a livestock producer can only set target prices by knowing actual or accurately estimated production costs. A marketer must also know what the market is paying, or is expected to pay. The level and timing of these target prices should be set based on quality market outlook information, cost of production figures and cash flow needs rather than expected profit levels. The advantage of setting several target prices rather than just one price allows a producer room to respond to changing market trends. Staying in touch with the market is crucial when trying to hit a target price.

Evaluate Pricing and Delivery Alternatives

Producers should evaluate all available alternatives for pricing and delivering their livestock. Each alternative has specific features that may make it more suitable than another in certain circumstances.

There are several livestock pricing choices available for any strategy. A forward contract offers a producer an opportunity to lock in a price for his livestock ahead of an expected sale date. Other alternatives are available for pricing livestock. They include open bids at auction markets, producer or breed association sales, video auctions, satellite and internet auctions, direct sales to packers, sales to livestock order buyers or using the futures market and a hedging strategy. Producers should keep their target prices in mind as they consider each pricing alternative.

There are a variety of pricing methods for the market delivery alternatives listed above. These pricing methods determine such things as whether an animal is sold live or rail graded and whether it is sold with a pencil shrink or not. All aspects of a pricing agreement will have a direct influence on the final return a producer receives.

When evaluating marketing alternatives, producers should keep in mind how their animals will be delivered to the buyer and if this delivery method will influence the settlement price. The method of transport includes both the operating costs of the truck and the costs of lost weight or quality of the animals. These factors should be considered as producers decide how and where to have their livestock priced. Pricing and delivery decisions are typically made together when selling. The pricing decision will sometimes dictate what the delivery method will be. However, both pricing and delivery methods can often be negotiated when reaching a settlement price with a buyer. A producer, who knows production costs and cash flow needs, can better determine if the price being negotiated is suitable for the producer's business needs or personal profit goals. A producer, who knows current market conditions, can better determine if an offered price is reasonable for current conditions.

Stick to Your Plan

A livestock marketing plan involves all the steps listed above. Producers, who follow these steps, will have a thorough understanding of how their business is functioning. They will also have the confidence to stick to their plan as they watch the market change daily. Changing plans on the spur of the moment can be as bad as having no plan at all.

Evaluate Your Plan

All plans must be evaluated to determine what worked and what needs be improved on in the future. The need for evaluation also applies to the seven-step marketing plan. A producer, who looks back on livestock sales and how the returns received matched the needs of a business, will continue to learn more about what factors influence his operation. This learning process will provide opportunity for growth in the future. Marketing is more than just selling. For your farm to be a successful business, it must include marketing as part of the overall farm management operation.

Why Study Livestock Markets?

As already stated, in most African countries there is a severe paucity of time series data on livestock prices as well as on the performance and efficiency of the livestock marketing system. Ironically, livestock marketing happens to be a favourite sector, where African governments choose to intervene in a variety of ways. These interventions range from outright fixing of wholesale and retail meat (e.g. Benin, Ethiopia, Togo) to monopolising the export market (e.g. Botswana, Kenya). Yet in many instances policy decisions on livestock marketing are taken in the absence of vital information on how they affect livestock producers, traders, slaughter-houses, butchers and consumers. Very often price fixing at unrealistic levels leads to open black markets, where the real prices substantially differ from those officially listed. In spite of this the official prices constitute the price series data, which clearly distort any analysis based on them.

Even when governments pursue price stabilisation policies it is difficult, in the absence of livestock market data, to establish to what degree their effects are transmitted to the level of producers. The absence of data on the magnitude and seasonality of supply as well as prices can frustrate the success of development projects. The closure of the meat packing plant at Kotsi in Sudan after a few months of operation is a case in point (Abbot, 1979). It is often argued that the stratification of the beef industry into areas of breeding, growing out and fattening should be pursued; this is often included in livestock development projects without much success. The comparative advantage of such a proposal cannot be fully assessed without determining the long-term stable price margins between the areas of stratification. An important use of time series data is in assessing over time the terms of trade between livestock producers and the rest of the economy. This assessment can be made using weighted price indices of a pastoralists' consumption basket and comparing it to the weighted price indices of their sales basket (Swift, 1979). Such an analysis should be made from time to time to gauge and temper the effect of price policies on pastoralists, who depend on the market for subsistence much more than agricultural households.

Thus in any country, livestock marketing studies are essential to provide vital information on the operations and efficiency of the livestock marketing system for effective research, planning and policy formulation in the livestock sector.

A Livestock Marketing System Model

A schematic representation of a livestock marketing system, the bottom part shows the flow of livestock from producers to secondary (regional) and terminal (national) markets through one or more primary collection markets. Livestock markets can easily be differentiated by the type of sellers and buyers operating in the market and the purpose for which livestock are purchased.

The external and internal factors that influence the livestock marketing system. First on the supply side the cash needs of producers, the strength of demand for their livestock, and pastoralists' expectation of the nature and length of the dry and wet seasons influence the volume of the different species of livestock on offer at any time. The higher the cash needs of the pastoralists the greater the volume of livestock on offer. Their response to market demand has been a subject of controversy in the literature. However, there is growing evidence that pastoralists in fact dispose of their marketable animals in a manner consistent with sound economic behaviour (Ariza Nino et al, 1980). In other words the stronger the effective market demand as expressed by high prices, the greater the volume of livestock supplied. Finally, pastoralists' perception of the climate influences supply and hence the price of livestock. Anticipation and occurrence of prolonged dry seasons induce more sales. The poor condition of the animals plus the greater numbers supplied during such times depress livestock prices. On the other hand the anticipation and occurrence of good raids causes pastoralists to withhold animals from the market so that they can put more weight and fetch better prices later on.

Second, government policy through fiscal, regulatory and development intervention affects the volume, flow and prices of livestock in the marketing system. Favourable fiscal policies that encourage livestock production and reduce costs to producers increase the supply of livestock, e.g. subsidies, and price stabilisation policies. On the other hand taxes and levies of all kinds tend to restrict the volume supplied. The control of epidemic diseases, the proper development of range areas and the development of trek routes and livestock market facilities tend to increase the volume supplied and reduce marketing costs. In general government monopolistic tendencies and the fixing of artificially low prices stifle market supply and demand.

Finally, market demand as expressed by the volume and prices buyers are willing to pay for livestock influences the behaviour of the markets at all links in the system. The efficiency of the market as reflected by the marketing costs of the system and to what extent price changes are transmitted through the marketing system strongly influence the operation of the markets. The less efficient the market the less responsive will supply be to changes in market demand.

Livestock Market Research Methodology

Livestock marketing systems research involves a two-pronged approach. The first involves the regular collection of time series data from a network of selected livestock markets. The second deals with in-depth studies of the performance efficiency of the livestock marketing system at the various links of the chain as livestock move from producers to the consumers.

Identification and Selection of Livestock Markets

The first consideration in setting up a livestock market data collection network is the identification of livestock markets and the selection of those markets to be included in the network.

Livestock markets in each administrative unit of a country, classified in the manner described in previous section, can be put on a map with arrows indicating the direction of flow of livestock by species. Extension staff can be effectively used to provide information for such classification and estimating roughly the volume of different species of livestock offered for sale on each market day.

The selection of which markets to include in the network, like any sampling problem, is a function of the coverage desired and the resources available for collecting the information. If there are several primary collection markets feeding a secondary redistribution market, it is not necessary to regularly collect data from all of them. A few can be selected on the basis of location, distance from the secondary market and the volume of supply. Seasonal observations can then be used to correlate those left out from the network with the secondary market in order to estimate their supply and prices.

Types of Livestock Market Data to be Collected

The volume of livestock on offer at the market is important as it gives a picture of the supply as well as its influence on prices. If possible the volume should be recorded by species, sex and age. Such information is easier to obtain in fenced markets where there are controlled gates. It may be difficult in open areas where livestock are bunched together in mobs. In such cases number by species is sufficient. The sex and age structure can be estimated by sampling the mobs seasonally.

The number of sellers and buyers participating in the market should be recorded to indicate the degree of concentration and hence influence on the price of livestock and offtake.

The price of livestock transacted should be recorded by breed, sex and age either on a census or sampling basis depending on the volume and the number of enumerators available. For large markets where there are more than 200 animals sold per species a 25-30% sample is desirable. In a market where one expects two breeds and their crosses this yields a minimum of three to six observations per class of animal by breed, sex and age.

A survey of prices producers receive should be made periodically to establish their relationships with market prices.

The weight of traded animals in the sample if possible, should be recorded. However, it may be difficult to weigh livestock in unorganised markets, where there is no auctioning on a weight and grade basis. In such cases visual assessment of body size (large, medium, small) and body condition (good, fair, poor) coupled with the age category of the animal can give a good indication. Weight correlations with body size and body condition can be established from the nearest abbatoirs.

The destination and purpose of traded animals should be recorded as such information will give a good picture of the direction and magnitude of flow and the purpose for which animals are purchased.

The mode of transport should be recorded as it will show the importance of trekking, railing and trucking over time.

All of the above information should be recorded by field officers or enumerators from the selected markets on the market days predetermined by the market research officers. In addition to this they should comment on conditions that influence the supply and demand situation for that market day.

Frequency of Data Collection

Once the livestock markets from which data will be regularly collected are determined, the next question is at what frequency should the data be collected? This depends on the type of livestock market in question and the frequency of market days. In primary livestock markets livestock are usually traded once or twice a week. Market days of secondary or regional livestock markets do occur more frequently. The frequency of a terminal market-day may be as much as six times a week.

It is desirable for time series analysis to have weekly data but it is not necessary to collect data at each market day. It can be collected on a sampling basis after establishing the representativeness of the market days with respect to the week. Are all market days held during the week similar or are some market days more important than others in terms of volume offered and volume traded? For instance if in a secondary livestock market both Wednesdays and Fridays are market days and Fridays tend to be more important, one can establish the factor (X) by which the volume supplied and the volume traded is greater than on Wednesdays by conducting an initial survey of the market as well as cross-checking the information by interviewing buyers and sellers in the market. This factor can then be used to estimate the market parameters that pertain to the Wednesday markets from data collected on Friday markets.

In livestock markets where there is only one market day per week a similar approach can be taken in sampling one week or two weeks per month. However, one has to be cautious and take into account that seasonal changes will be captured by such sampling.

In many African countries (e.g. Ethiopia and Nigeria) the occurrence of major religious (Christian and Muslim) holidays has a market effect on supply? demand and prices of livestock, especially those of smallstock. Demand is high during these holidays and prices can be 80% more than the annual average price (Okali and Obi, 1982). It is therefore essential to intensify livestock market data collection during such holidays in order to accurately assess their impact on the various market parameters.

Studies of Market Performance and Efficiency

Studies of the performance and efficiency of the livestock marketing system at the various links of the chain as livestock move from producers to consumers (including the wholesale and retail trade of meat) have to be conducted by a senior livestock market analyst who is in charge of the entire livestock market data collection and analysis with assistance in the field by his colleagues at headquarters and the provincial livestock marketing officers. The reason for this is twofold. First the observations and probings necessary to get the information require a high degree of skill and experience. Secondly, the senior market analyst has to acquire a first-hand insight into how the livestock market operates in order to properly analyse and interpret the time series data being collected by the field officers.

Studies of market performance and efficiency include two major aspects of the livestock marketing system. The first is an assessment of the degree of buyer concentration in the markets selected for time series data collection and how livestock prices are arrived at and purchasing is financed. Although the number of buyers and how many animals they bought in the market can be recorded by field enumerators, assessing the manner in which they operate, whom they represent, in how many other livestock markets they trade, and how they finance their purchase requires a considerable degree of skill and market knowledge to elicit.

The second study involves establishing the cost of livestock and meat marketing as animals change hands from the producer to the primary markets, to the secondary markets and finally to the terminal markets, where they are slaughtered for domestic consumption and/or are exported live.

These marketing costs can be distinguished as costs of:

(i) transporting (trekking, trucking and/or railing);

(ii) feeding (including grazing);

(iii) marketing levies and taxes imposed by local and national authorities;

(iv) mortality or loss (some animals die during transit because of diseases or other physical stress; some might stray and not be recovered);

(v) slaughtering and processing costs;

(vi) capital as represented by the interest on the money tied up by the livestock from the point of purchase to the point of sale; and

(vii) the opportunity cost or salary of the operator (trader, butcher etc.).

The above information can be established by interviewing livestock traders and managers of slaughterhouses and spot checking the information by actual observation on their operations. Livestock marketing margins can be defined as the difference between the sales price of the animal (meat) and the costs incurred by the seller including the acquisition price of the animal (meat). The less the margins the more efficient the marketing system.

Analysis of Livestock Market Data

The types of analysis of livestock market data are categorised under four major headings:

(i) supply of livestock;

(ii) destination of livestock;

(iii) price movements; and

(iv) market performance and efficiency.

Supply of Livestock

The sources of livestock supplying the particular market can be analysed using frequencies of traded animals by area of origin by breed, sex, age and total volume. This indicates what types of animals are supplied by each area and the frequencies can be used to estimate marketed offtake from the hinterland of the market.

A problem which frequently arises in such analysis is to what detail origins of livestock need to be specified. One can *a priori* section the hinterland by a functional criterion (geographic, type of producer or source) and instruct enumerators to categorise origins of livestock into the specified sections. Alternatively, one can instruct them to record place names of the origins of livestock and decide later how to categorise them. Table blow shows an example of such an analysis by type of producers.

Table: *Source of cattle supply to the Emali market*

Source	***No. of cattle***	*%*
Group ranches in Kajiado district	1143	77.9
Trading centres serving group ranches in Kajiado district	292	19.9
Individual, ranches Kajiado district	23	1.6
Farms in Machakos district	9	0.6
Total	1467	100

Source: Bekure et al (1982).

Destination of Traded Livestock

The frequencies of traded livestock by breed, sex, age and purpose of purchase for major destinations reveals the relationships between various livestock markets as well as the trade in livestock between pastoralists, agro-pastoralists and agriculturalists. The total volume and fluctuation of livestock supply are of major interest as they show the degree of seasonality of supply.

This can be easily seen by plotting weekly or monthly supplies. For long-term determination of seasonality several years' time series data is essential. A standard statistical technique for estalishing seasonality is the method of moving averages. The most important single parameter in collecting time series data on livestock marketing is livestock price. For a given market and a given period (week, month, year) mean livestock prices per head and liveweight can be analysed by breed, sex and age. The next analysis is that of determining whether there are seasonalities in livestock prices. Seasonality of prices can be determined by type of animals sold. The methodology is similar to that of determining seasonality in supply of livestock. However, several years' data are required to establish long-term seasonality.

Finally, a stepwise multiple regression analysis can be used to fit a demand model for a given market relating price to breed, sex, age, season, volume of supply and number of buyers in the market as well as the interactions between these independent variables (Drapper and Smith, 1966). The longer the time series data the better the specification of the demand function. The value of the demand function does not lie so much in its ability to predict future prices but in its usefulness in quantifying the relationships between livestock prices and breed, sex, age, season of the year, type of market and other variables one may specify. Shapiro (1979) analysed price of cattle in Upper Volta as a function of age, sex, season of the year, region of the market, type of seller etc. The results of the regression are summarised as follows:

- Prices for males increase at an increasing rate with age, up to 5.7 years (average 5,000 CFA F per year overall); they increase at a decreasing rate to age 11.4, where they begin to fall.
- No strong age-price relation was found for females.
- A premium of 1,500 CFA F was paid for steers over bulls at all ages.
- Only slight evidence of higher prices was found for sales closer to major consumption and export centres.
- No difference was found between ethnic groups as sellers.
- Higher prices occur during the rainy season, lowest during the dry season; males hold their prices better than females during the dry season.
- The amount of seller market information had no significant effect on prices.
- Higher prices were paid for males when they were sold to butchers and traders; females brought higher prices when sold to herders and farmers.
- The type of market had no significant effect on prices.

The foregoing analysis on time series data of livestock markets can be done on district, province and national levels to give information at various levels of aggregation.

Market Performance and Efficiency

Analysis of the studies of the operations of livestock markets yield in the first instance qualitative and quantitative information on the Operations of the markets that is useful in analysing and interpreting the time series data generated. There is no set way of analysing such data. Studies of the well established traditional marketing systems in West Africa, which also deal with exporting live cattle from the northern pastoral areas to the coastal zone, show that they perform efficiently (with gross margins of 1520%) despite their traditional base and complexity (Herman, 1979; Staatz, 1979). Our own work (Bekure et al, 1982) and that of Matthes (1979) in Kenya show that marketing margins are much higher (25-35%).

Organisation for Collection and Analysis of Livestock Market Data

There cannot be a set way of organising a national livestock marketing data collection and analysis. One can only make a general suggestion which elaborates the main features. In the end the organisation to be adopted in a given country will have to take into account the prevailing conditions. What is suggested here is a flexible organisational framework which can be easily adopted with modification. A schematic representation of a general organisational framework for the collection and analysis of livestock market data. It is based on the hierarchial structure most common to ministries of livestock, or agriculture in many African countries. Their field offices are usually hierarchially organized in conformity with administrative units of their respective countries. Thus, the ministries have field offices at provincial, district and occasionaly at locational levels, where field officers with secondary school education are posted.

It is suggested that the organisation for the collection and supervision of livestock market data follows the same structure. At the headquarters of the ministry a Livestock Marketing Analysis Section should be responsible for organising the collection, supervising and analysing all data on the livestock market system of the country.

This section should be located within the Livestock Marketing Department so that it falls under the responsibility of a department with a functional commitment to the task. In ministries where there is no such department it may be placed under the Planning Unit. The section should be headed by a competent senior livestock market analyst with access to a statistician and data processing facilities. How big and permanent the staffing of the section should be is a function of the size and the importance of the livestock sector in the country. It should therefore be tailored to the needs of the country in question.

At the provincial level, it is suggested that a full-time junior market analyst be made responsible for supervising and partially analysing the data collected in the province. This will enable prompt supervision of data collection as well as facilitate feed-back of market data to market participants at the local level. For instance average weekly prices of important categories of livestock in major district markets can be compiled at the provincial office and promptly reported to the public by the mass media. This in itself may increase the efficiency of the livestock marketing system.

In addition to supervising and ensuring good quality data collection at the district level, the provincial livestock market analyst will be responsible for conducting in-depth livestock market studies under the direction and supervision of the senior livestock market analyst at the headquarters of the ministry.

At the district level, it is suggested that one of the already existing personnel of the ministry be made responsible for supervising data collection and passing the information to the provincial livestock market analyst. In many African countries district level personnel are university graduates with sufficient background to handle the task provided they are given adequate orientation and on the job training. In cases where the district is located in a major livestock trading region a full-time supervisor may be warranted.

At the market level it is our contention that most of the time series data collection on livestock markets can be conducted and supervised by existing personnel, who in many instances have offices a few yards from these markets. In case livestock markets, where no ministry officers are posted, are selected, other enumerators (e.g. teachers, bussinessmen etc.) may be contracted on a part-time basis, provided they are adequately supervised with unannounced spot checking. We have used this approach at Ong'ata Rongai, Kenya with good success.

In conclusion, we would like to reiterate that the above organisation is suggested only as a framework for consideration. In small countries or in countries where the network of livestock markets is small and caters for only one terminal market, as in Botswana, the whole organisation may mean one senior and two junior analysts being assisted by field enumerators.

Chapter 3

Management of Livestock Production

Fibre digestibility: *Streptococccus bovis*, a numerically predominant bacteria in the rumen of crossbred cattle under different feeding regimes was charaterized. And the cellulase gene obtained from the best fibre degrading fungi isolated from the faecal matter of goat was cloned to *Streptococccus bovis* for enhancing the fibre digestibility of poor quality crop residues. Suppression of methane production: *In vitro* gas production test revealed that methane suppression among the tree leaves ranged from 4.6 to 82%— minimum recorded in jatropha leaves (4.6%) that contained lowest tannin and the maximum (82%) in *Ficus bengalensis* leaves.

Commonly used top feeds like *Sesbania grandiflora*, *Glyricedia maculata*, *Ficus mysorensis* and *Ficus religiosa* leaves showed a methane suppression ranging between 35 and 50%. A mixture of three plant species having antimethanogenic activity *in vitro*, exhibited 12% reduction in methane emission in crossbred calves, which confirmed that there is a potential in using tree leaves for reducing methane production from enteric fermentation.

Use of fungi for enhanced digestibility of straws: Lignin content of *ragi* straw decreased with all the white rot fungi, viz. *Phanerochate chrysosporium, Pleurotus sajorcaju, Pleuritous ostreatus* and *Voriалla voloraceae*. High protease activity was observed during the first two days of fermentation after which it declined, and lignolytic enzymes, viz. laccase, manganese peroxidase and lignin peroxidase concomitantly increased up to the fifth day of fermentation. *Phanerochaete chrysosporium* showed the best potential in improving the digestibility of *ragi* straw followed by *P. ostreatus*. Antifungal property of medicinal/

aromatic plants: Melissa and thyme leaves exhibited good anti-fungal activity (>50%) against *Aspergillus parasiticus*. *Pachouli* leaves, curry leaves, rosemary leaves, cinnamon leaves, *sarpagandha* leaves, thumbe leaves, sweet worm wood leaves, *aswagandha* leaves, chicory powder, yellow oleander leaves, *Selastras paniculatus*, *Tinospora cardifolia*, railway creeper and Indian acalypha leaves showed high anti-fungal activity (>50-90%) against *Fusarium moniliforme.*

Buffalo

Nutrition for the onset of puberty: GnRH challenge studies in the calf attaining maturity at 3 years of age due to nutritional perturbation showed immature status of hypothalamohypophyseal-ovarian axis even at the age of 2 years 4 months. Nutritional modulation of IGF: In sub-fertile male buffaloes nutritional modulation of IGF-I (as a mediator of metabolic hormonal effects) proved beneficial for various sperm functional attributes *in vitro*. The cleavage rate in embryos produced through IVF using such sperms, was also better.

Area Specific Mineral Mixture: Supplementation of area specific mineral mixture (Ca, P, Zn, Mn, Cu) to buffaloes based on the deficiency in the North east zone of Haryana, improved productive and reproductive efficiency as 70% buffaloes showed normal cyclicity, and 10% increase in milk production. Supplementation of area specific mineral mixture pellets @ 40g daily in cattle and buffaloes during lactation stage increased milk yield by 10-15%, brought cows into estrus within 30-45 days, and reduced the problem of skin keratinization.

Detection of pesticide residues: A rapid multiresidue method for analysis of neonicotinoid pesticides, viz. imidacloprid, acetamiprid and thiacloprid, was developed. The percent recovery from 0.5 to 2.0 ppm concentration varied from 95.6 to 81.17% for imidacloprid, 92.76 to 84.99% for acetamiprid, and 96.96 to 88.50% for thiacloprid with a detection limit of 5ppb, 10ppb and 20ppb, respectively.

Fibre digestibility: Isolates of anaerobic fungi collected from 5 different states and 6 different host species, revealed that isolates from Rajasthan had highest fibrolytic potential and the best isolate improved digestion of wheat straw in buffaloes.

Improving Eproductive Efficiency

Buffalo embryos were produced *in vitro* @ 64% cleavage with 20% blastocyst using an improved IVEP protocol.

- A three dimensional (3D) collagen gel culture system for the *in vitro* growth and survival of the buffalo preantral follicles with IGFI was developed.

- Early diagnosis of pregnancy (by day 20-21 post-breeding) was facilitated by using real time B-mode ultrasonography in goats and buffaloes.
- Bull-biostimulation curtailed the incidence of silent ovulation and service period and increased conception rate in post-partum buffaloes.

Enhancing Productivity: Strategic supplementation of protein (<20% CP) during midlactation increased productivity of buffaloes.

Vitamin E supplementation @300 IU daily was optimum to improve weight gain, and increased its concentration in the muscles of buffaloes. Early embryonic mortality: Dynamic status of antioxidant enzymes in relation to the stages of oestrous cycle and tissue remodeling was observed. Effective modulation of prostaglandin production by the uterus may rescue corpus luteum and prevent early embryonic mortality. Improvement of reproductive efficiency: Insemination dose could be reduced from 25 to 15 million spermatozoa without adversely affecting the conception rate. The results are being authenticated with more trials in farmers' herd. Frozen semen samples were evaluated for sperm motility attributes. Bulls with higher field conception rates also had higher sperm total motility, progressive motility, rapid motion and viability.

Sheep

Trace Element Status: Biochemical markers (Cu and Zn-dependent enzymes – ceruloplasmin and Cu/Zn- super oxide dismutase) were evaluated to assess the trace element (copper and zinc) status in sheep at different dietary levels of Cu and Zn. Prediction equations developed by correlating the absorbed Cu and Zn with Cu- and Zn-dependent enzymes; plasma Cu and Zn with Cu and Zndependent enzymes; liver tissue concentrations of Cu and Zn with absorbed Cu and Zn, revealed that Cu and Zn status of animals could be assessed by using these enzymes as biochemical markers. Production performance: Bharat Merino and Gaddi Synthetic sheep under migration to highland pastures gained higher body weights and produced more wool in comparison to stationary flocks.

Utilization of Fibrous Crop Residues: Fortification with cellulase, xylanases, pectinase, phytase and protease enzymes enhanced dry matter digestibility by 7.00%, cell wall digestibility by 25.00%, and the end product fermentation by 15.00% of poor quality roughages. Supplementing probiotics of microbial origin like *Saccharomyces cerevisiae*, *Saccharomyces uvarum* and *Kluyveromyces marximanus* and a mixed yeast culture of above three in a ratio of 1:1:1 as microbial probiotics in lambs showed that *Saccharomyces cerevisiae* strain is superior in improving the growth of lambs. Bioavailability of nutrients: Diets supplemented with condensed tannins improved nutrient utilization, immune response besides protection from GI parasites and fasciolosis in sheep.Supplementation of *Tinospora cordifolia* as a functional food imparted positive influence on the nitrogen metabolism and antioxidant levels in seminal plasma besides significantly improving the erythrocytic antioxidant status and cell-mediated immune response of adult Muzzafarnagari rams. Undecorticated jatropha (*Jatropha curcas*) meal after processing with 1% common salt and 0.5% lime, could replace protein of conventional oil cakes up to 25% in the concentrate mixture of adult sheep and goat for short-term feeding, Goat.

Standardization of in Vitro Fertility Test

Hypo Osmotic Swelling Test (HOST): Sperms were evaluated for strongly coiled, weakly coiled and non-coiled under oil immersion lens. The best swelling in terms of strong coiling and total coiling was in 75-mosmol hypo- osmotic solution. There was significant difference in swelling in different strengths of hypo osmotic solution. For frozen sperm 75-mosmol and for fresh diluted semen 100-mosmol hypo-osmotic solution was found to be the best.

Dual Staining Test: Dual staining technique was standardized for testing viability and acrosomal integrity in frozen and fresh semen. It saved time and chemicals in testing viability and acrosomal integrity thereby avoiding separate tests.

Semen Quality: Twice a week semen collection, evaluation and freezing in Jamunapari bucks of 2-5 years of age group, indicated that the semen production was higher under intensively reared bucks compared to the semi-intensively managed bucks.

Rapid Estrus Detection Methods: Sponges of different sizes and shapes were tested for their retention in vagina for 12 days in Sirohi goats for estrus detection. Circular sponges with a diameter of 30 mm and cylindrical shaped sponges with a diameter 25 mm had the highest percentages (>83%) of retention. Mithun Feed blocks with locally available feed resources: Feeding of *Lagerstroemia speciosa* tree leaves based complete feed blocks to mithuns showed that the tree leaves could be incorporated in the ration up to 30% for feeding mithuns under semi-intensive or intensive system.

Organic Fertilizer: The excreta (faeces and urine) of both mithun and Tho Tho cattle were compared as a source of organic fertilizer.

The quantum of faeces voided from mithun was more than that of Tho Tho cattle though faeces of Tho Tho cattle contained less water (more DM) compared to that of mithun. The chemical composition of faeces of mithun and local cattle did not differ significantly. Mithun produced more urine than Tho-Tho cattle. So per animal basis, mithun supplied more excreta as organic fertilizer than Tho-Tho cattle in Nagaland. Bakers yeast: Feeding bakers yeast (*Sacharomyces cerevisiae*), a probiotic on roughage based diet, significantly increased average daily gain of mithuns, intake of concentrate and roughage and also improved FCR.

Estrus Synchronization Protocols: Experiments conducted to synchronize estrus in cyclic and postpartum mithun cows showed more prominent behavioural signs of estrus than spontaneous heat. Application of CIDR on day 45-50 after parturition, induced first postpartum estrus immediately after uterine involution (day 53-58 post parturition). Unlike other bovines, mithun cows exhibit first postpartum estrus at around 97±19.6 days postpartum. Use of CIDR was advantageous in terms of prominent behavioural signs of estrus thereby ease in detection of estrus. The first calf was born from an anoestrus mithun cow synchronized with CIDR. Hormone-induced maternal behaviour: Mother-neonate bonding was studied using oxytocin intranasal spray. Intranasal administration of oxytocin effectively induced maternal behaviour in primiparous bovine heifers where maternal behaviour was blocked chemically.

Embryo Transfer Technology: Estrus synchronization was performed by using CIDR protocol and four embryos (compact morula) were recovered successfully from two donors and subsequently transferred into three recipients.

Trace Mineral Supplementation: Soil, feeds and fodders of yak rearing zones are deficient in micronutrients, as reflected by the low productive and reproductive performances of yak. Supplementation of trace minerals like Zn, Cu, Co and Mn in the ratio of 40:20:2:1 along with the basal diet significantly increased milk production.

Production Performance During Winter: Body weight gain was significantly higher during October in calves but from November onwards no significant increase was observed. Bulls gained significantly higher body weight compared to calves up to December. The lactating yak cows lost about 5.84% of their body weight, and milk yield reduced mainly due to shortage of feed and fodder during long winter. Providing adequate nutrition during winter could help in ameliorating winter stress in yaks.

Modified Temperature Humidity Index: Yaks were comfortable at THI of 52 and when THI exceeds 52 yaks experienced heat stress, as expressed through increased physiological responses.

Poultry

Stress Related Hormone: Under heat stress condition some of the lymphocyte proteins were repressed whereas some others were induced in broiler.

Bioavailability of Micronutrients: Se supplementation in broiler chicken diets at 0.15 or 0.30 ppm complemented bioavailability of Zn. In contrast, Se antagonized retention of Mn, Cu and Fe in liver tissues. Se (0.15 ppm) and Zn (80 ppm) improved humoral and cell-mediated immune response in broiler chicks. Zn uptake by tissues was relatively more active during early age (2 week) than at later ages (4 and 6 week), whereas Se retention in tibia and liver was higher at 4 and 6 weeks than that at 2 weeks of age. Se inclusion in broiler chick diets from 0.15 to 1.35 ppm linearly enhanced its retention in bone and liver tissues and complemented Zn uptake by tissues. Both Cu and Fe responded negatively to Se increases in diets at 5 weeks of age. Vitamin E at 40 IU enhanced Se uptake by bone, but did not influence retention of Zn, Mn, Cu or Fe. Moderate levels of Se (0.15 or 0.45 ppm) and vitamin E at 40 IU produced higher antibody titres, better cell-mediated immune and reduced stress in 5-week-old broilers.

Enhancing utilization of macronutrients: Protease enzyme produced from *Bacillus licheniformis* were supplemented to broiler diets @ 4,000 IU/kg to enhance the feeding value of commercial meat meals and soybean meal low in protein by 3-4% over the recommended level (22%). Meat meal diets responded better to protease supplementation and performance of broilers was equivalent to the control group that was maintained on 22% protein diet. Inclusion of enzyme had significant impact on different production parameters compared to non-supplemented diets, particularly when dietary protein levels were lower than the recommended levels.

Female Reproductive System of Desi Fowls: Ovary and oviduct development was noticed clearly around 16 to 18 weeks of age in White Leghorn (WLH), around 20-22 weeks of age in Kadaknath (KN) and around 24 weeks in Aseel peela (AP). At the peak of sexual maturity, around 30 weeks of age, total length of the oviduct was greater in WLH (73 cm) as compared to *desi* fowl (66 cm). The transaminases activity of blood plasma, irrespective of breeds increased linearly with age. An increased pattern of ACP, GOT and GPT activity was found associated with maturation of female reproductive tract and reverse was true with ALP activity among all the breeds.

Enhancing Egg Production: Using simple feed formulation, egg production could be enhanced markedly over the age of 78 weeks in Aseel peela *desi* fowl. Large-scale replications at institute and field level are being taken up to validate the data.

Supplementation of Melatonin: Dietary inclusion of aflatoxin @0.15 ppm level adversely affected body weight, feed intake and FCR and caused lipid peroxidation with simultaneous depletion of antioxidant enzymes (superoxide dismutase and catalase) in broilers. Melatonin supplementation @ 40mg/kg feed alleviated the adverse effect of aflatoxicosis at lower levels (<0.15ppm).

Moulting for enhanced production: Birds were force moulted by feed withdrawal method for 10 days. As the period of fasting progressed from 0 to 10 days a gradual but steady reduction was noticed in the levels of serum triglycerides, which was more pronounced and significant from the fourth day of commencement of feed withdrawal.

A similar decline in serum total cholesterol concentration was noticed, which became very apparent from the sixth day after feed withdrawal and a reduction of 30% in total cholesterol concentration was achieved by the tenth day in the moulted hens. In contrast, serum HDL– cholesterol concentration progressively increased during the feed withdrawal period and peaked around the eighth day after withdrawing feed. Accumulation of high lipid in uterus in late laying age results in either shell less or poor shelled eggs. Feed withdrawal for longer period leads to mobilization of lipids from uterus.

Quail Semen Characterisation: Physical and biochemical characteristics of quail semen showed that birds having larger cloacal gland size ejaculated higher volume of semen and semen production was higher in CARI Uttam than CARI Sweta quails. Mass sperm motility in neat semen was only 50-60% immediately after collection that decreased continuously and reached to zero after 30 min.

Sperm abnormalities were higher in CARI Sweta than CARI Uttam. Among the enzymes LDH was exceptionally high in all groups in both the lines, cations sodium and potassium were higher in birds having larger cloacal gland size whereas magnesium and calcium were more in the seminal plasma of birds with smaller cloacal gland. Methylene blue reduction time test revealed that quail spermatozoa are more active than chicken.

Livestock Protection

A status of freedom from contagious bovine pleuropneumonia infection in cattle and buffalo was obtained from OIE.

Development and Improvement of Diagnostics and Vaccines

Vaccines

- A low volume saponified haemorrhagic septicaemia vaccine was found safe and effective in farm cattle.
- Possibility of DNA vaccine construct against bovine brucellosis was ascertained.
- Chicken cytokine genes (MIP-b, lymphotactin and IFNg) were expressed in mammalian cells thereby opening up the possibility of their use as genetic adjuvant in DNA vaccine.
- LPS and genomic DNA containing CpG from *Salmonella* Gallinarum activated the innate immune system of chickens and gave higher protection after immunization with inactivated NDV in challenged birds.
- Conjugation of Fc with flagellin protein was a good model for efficient antigen delivery resulting in higher immune response than antigen alone.

Diagnostics

- A nested RT-PCR was developed using primers from RNA dependent RNA polymerase region for differentiation of ruminant pestiviruses.
- C18L gene based conventional PCR and TaqMan probe based real time PCR were developed for specific detection of buffalo pox virus.
- Duplex PCRs were developed for specific detection and differentiation of buffalo poxvirus from other orthopox viruses, and camel pox from other orthopox viruses.
- C18L gene-based real time PCR was standardized for quantification of camel poxvirus in clinical samples.
- A hybridizing probe based real-time PCR was developed for diagnosis of PMWS and the disease was diagnosed in four private farms in Uttar Pradesh.
- The expressed protein of N gene of PPR virus could be an alternative to whole virus antigen in sandwich ELISA for diagnosis of PPR.
- Developed indirect ELISA for serodiagnosis of Japanese encephalitis in pigs.
- Transformed fibroblast antigen was much superior antigen for detection of anti-avian leucosis virus antibodies in the serum

samples of chicken, as compared to gsAg, as determined by an indirect ELISA.

- A highly sensitive PCR targeting new gene of *Mycobacterium a. paratuberculosis* and a quantitative real-time PCR (RT-PCR) were developed for the diagnosis of paratuberculosis in small ruminants.
- Serotype specific PCR was developed for detection of *Salmonella* Typhimurium and *S.* Enteritidis.
- Duplex PCR was developed for simultaneous detection of *Salmonella* genus and Typhimurium serotype.
- Germ tube formation test was developed for detection of chlamydospore in *Candida* albicans.
- Methodology for quick detection of *Echinococcus granulosus* genotypes by polymerase chain reaction coupled with restriction fragment length polymorphism was developed.
- A useful primer was developed and found effective in differentiating cryptic stage of *Echinococcus granulosus* and *Taenia.*

Molecular characterization of pathogens/receptors:

- Mutants of *E. granulosus* isolate from Indian cattle and buffalo origin, were detected on the basis of sequence analysis of mitochondrial gene and non-coding spacer gene.
- Molecular characterization of toll-like receptors (TLR2, and TLR4) of *nilgai* revealed higher expression in skin and immune cells of *nilgai*, as compared to buffalo indicating stronger innate immunity.

Herbal Medicines

- Immunomodulators prepared with extract of *Tinospora cordifolia* and a probiotic (*Mycobacterium phlei*) showed significant body weight gain in broiler birds and improved their health.
- Significant antidiarrhoeal activity was detected in the seed extract of *Caeslipinea* bonducella.

Surgical and Clinical Interventions

- The epoxy-pin external skeletal fixation technique was developed, and used to treat a variety of compound fractures of different long bones in small animals.

- A novel design of bilateral external fixator having opposite threadings in the side bars was developed for the management of long bone fractures in large animals.
- Transplantation of autologous bone marrow cells, along with hydroxyapatite induced faster healing of radius fracture in rabbits, as compared to transplantation of hydroxyapatite alone.
- Application of autologous bone marrow cells subcutaneously in the periphery of incisional and open cutaneous wounds induced faster healing, as compared to conventional antiseptic dressing of wounds.

Foot-and-mouth Disease

Field samples (1,313) received from various states during the year were processed and subjected to sandwich ELISA for type identification. Only 705 samples were typed — 567 samples were typed as O, 58 samples as type Asia1, and 80 samples as A —, and no virus could be detected in rest of the samples. Samples were also processed in BHK 21 cells and virus could be recovered in 119 field samples comprising 24 type Asia 1, 75 type O and 21 type A. To improve the diagnosis of FMD in suspected clinical samples, a multiplex PCR (m-PCR) was developed and evaluated. Using the test, 42% of the outbreaks that went undiagnosed using ELISA, were identified indicating that mPGR could be used as best supplentary to ELISA to increase the percentage of FMD outbreak diagnosis in the country.

Two-dimensional micro-neutralization test (2DMNT), a modified form of SNT, was routinely used to test new field isolates to determine the appropriateness of the existing vaccine strains and to select new vaccine strains, if required. In serotype A the most worrying factor, which merits attention is the antigenic heterogeneity of the isolates. In the sense some strains show close antigenic match to the current vaccine strain (17/82) and others to the new strain (IND40/00) in *in-vitro* micro neutralization test. One isolate IND 53/08 from Chhattisgarh, was unique both antigenically and genetically forming a separate cluster with another isolate IND 109/06 from Chhattisgarh.

Among all serotypes prevalent in India, type A virus population is genetically and antigenically most heterogeneous in nature. VP1 coding (1D) region based molecular phylogeny has established circulation of four genotypes of type A so far in India. There is once again an upsurge in incidence of outbreaks due to lineage genotypes VII with amino cicer (aa) deletion at 59th position of VP3. This single aa deletion is at an antigenically critical position in structural protein VP3, which is considered to be a major evolutionary jump probably due to immune selection. Field isolates (17) of serotype A recovered from outbreaks in Karnataka, Tamil Nadu, Chhattisgarh, West Bengal and Haryana were sequenced at 1D (VP1) region for molecular epidemiological analysis.

The determined sequences were aligned with other Indian sequences and some of the retrieved exotic sequences. All the isolates clustered within genotype VII in the N-J tree. Genotype VII is restricted to only India as none of the exotic sequences clustered in this group. Thirteen out of the seventeen isolates sequenced, clustered in the deletion group. 1D region based phylogeny also revealed that this lineage is genetically diverging with time giving rise to three lineages (VII b, f and g) so far. In serotype O, PanAsia II strains dominated the outbreaks, nevertheless Panasia I and II 2001 also co-circulated. Asia 1 field isolates (19) were subjected to 1D gene sequence analysis. The isolates were grouped with lineage CI that dominated Asia 1 outbreaks. The isolates of 2007 and 2008 showed 15.4 to 16.7% and 12.4 to 14.7% divergence at nucleotide and amino acid level, respectively, from in-use vaccine strain (IND63/72). The Central laboratory, Mukteswar, contains 1,402 (893-O, 261-Asia 1, 233-A, 15-C) field isolates. Pre- and post-vaccinate serum samples collected up to sixth phase of FMD-CP showed increased levels of protective antibodies against serotypes O, A and Asia 1 over different phases of vaccination.

Animal Disease Monitoring and Surveillance

A large databank on the livestock diseases of the country, based on reports submitted to the Government of India by various state governments was developed at the PDADMS. The institution was involved in the sero-monitoring of rinderpest. Large number of sera samples from various parts of the country is maintained in the National Livestock Serum Bank for retrospective studies.

- An offline version of the databank of livestock diseases of the country was developed. Based on the custom queries, various epidemiological analyses are possible e.g., frequency of disease occurrence, top diseases of the country, eco-patho zones. The spatial and temporal analysis of animal disease data is being carried out using this software.
- Molecular diagnosis of brucellosis was standardized that helped in differential diagnosis of *Brucella abortus* and *B. suis*. Based on the results of the serological, biochemical and molecular techniques a rare case of brucellosis in swine due to *B. abortus* was diagnosed. A standardized A-B ELISA kit for the detection of bovine *Brucella* antibodies was developed. An indirect ELISA kit was standardized to identify the magnitude of disease in

ovines. Molecular epidemiological studies are being standardized to diagnose and differentiate the brucellosis of cattle, ovine, caprine and humans. Tests were developed to detect the etiological agent directly from the clinical samples such as aborted foetus, placenta and uterine discharges.

- A multiplex PCR was standardized to diagnose the pathogenic leptospira. A repository of the leptospira isolates is being maintained.
- Molecular studies on BHV-1 were carried out. A multiplex PCR for detection of BoHV-
- sequences was standardized. The PCR amplified products of gB (293 bp), gC (173 bp), gD (343 bp) and US 1 (464 bp) were subjected to partial nucleotide sequencing and aligned with different reference sequences of respective genomic regions.
- The PCR amplification of different 'tk'genomic region of BoHV-1 was standardized and the PCR amplicons thus obtained were confirmed using unique restriction enzyme. Multiple PCR using different combinations of primers specific for gB, gC and gD was standardized and was applied for screening of field samples. The partial nucleotide sequencing of gB, gD and US 1 were aligned with the reference sequences, and was analyzed with phylogenetic trees. These results would be of much help in profiling and characterizing BoHV-1 in livestock population.
- The serum samples obtained from Madhya Pradesh, Maharashtra, Andhra Pradesh, Manipur, Kerala, Orissa, West Bengal and Tamil Nadu, were screened for the presence of antibodies against IBR, using AB-ELISA kit and 41.90% of the samples were found positive for IBR antibodies.
- An mPCR for genome detection of leptospira, BoHV 1 and *Brucella* using known standards targeting the LipL32 gene of *Leptospira*, gB gene of BoHV 1 and bcsp31 gene of *Brucella,* was standardized.
- Occurrence of zoonotic bacterial pathogens from the livestock and livestock products was studied. The pathogens were isolated from various sources, and their molecular characterization was completed.
- A computer interface based BHV-1 whole antigen AB ELISA was developed as per the standards of IAEA, standardized and validated. The kit was critically evaluated both in-house and extensive field trials for detection of antibody to IBR virus in bovine serum. This test is highly sensitive, specific, economical and user friendly.

- A kit to detect the antibodies to *Brucella* in swine is being developed and is in the process of standardization.
- A PCR technique was developed to detect the carrier status in domesticated animals. A pair of primers specific to VSG gene of *Trypanosoma evansi* was developed. The PCR technique was standardized and 400 bp amplicon of VSG gene was obtained from the genomic DNA isolated from the blood of *T.evansi* infected experimental animal. Field validation of the technique is in progress.
- Serum Bank facility has more than 170,000 serum samples from all over the country, which is being used for long-term national surveys on various diseases of economic importance.
  * Development of relational database on Animal Health Information System
  * Development of *India.admasEpitrak* – a relational animal health information database software
  * Development and launching of National Animal Disease Referral Expert System
  * Identification of disease specific Eco-patho zones in the country
  * Providing eco-pathozones and spatial and temporal occurrence of diseases for effective vaccination and control of important diseases in different states e.g.

    PPR in Andhra Pradesh and Karnataka, Brucellosis in West Bengal and Andhra Pradesh, FMD, PPR, HS and BQ in Maharashtra, Andhra Pradesh and Karnataka.
  * Providing the logistic support to national network projects like bluetongue and HS projects for disease monitoring and surveillance.

Bluetongue

A repository of blue tongue virus isolates BTV- 1 (2 isolates), BTV-2 (4 isolates), BTV-9 (3 isolates), BTV-15 (5 isolates), BTV-18 (4 isolates) and BTV-23 (7 isolates) from Izatnagar, Hyderabad, Parbhani, Kolkata, Parbhani, Hisar and Chennai, was made. No outbreak of bluetongue was recorded in the country except Andhra Pradesh, Karnataka and Tamil Nadu. Disease forecasting model was developed. The incidence was as high as 95.5% sheep from Uttarakhand; 88.6% cattle from Panjab; 63.8% goat and 55.6% sheep from Manipur; 50.0% sheep from Jammu and Kashmir; and 18.22% goat from Delhi. A VP7 gene incorporated recombinant antigen based indirect ELISA kit was developed for detection of group specific antibodies in the sera. Inactivated pentavalent bluetongue vaccine was evaluated at different places particularly in the bluetongue affected states. Vaccine was satisfactory except a nodule formation at the site of inoculation. Type specific primer designing, VP2, VP5 and VP7 gene cloning and expression, multiplex RTPCR for BTV, RNA-PAGE and nucleotide sequence studies, were standardized. Confirmation of virus isolates was done by RT-PCR using VP7 gene specific primers.

Haemorrhagic septicaemia Isolates (93) of *Pasteurella multocida* were characterized and a new serogroup E of *Pasteurella multocida* was identified first time in the country. Most of the *Pasteurella multocida* isolates were sensitive to enrofloxacin, ofloxcin chloramphenicol, doxycycline and resistant to vancomycin, bacitracin, and sulphadiazine. Molecular characterization of different isolates of *Pasteurella multocida* recovered from different species of animals and poultry were carried out by PM-PCR, HSB-PCR, multiplex-PCR, ERIC-PCR and REPPCR.

A low volume saponified HS vaccine was validated successfully in farm cattle, and it was found satisfactory. The OMP vaccine against *P.multocida* type A in ducks provided higher protection as compared to the bacterins. The biofilm vaccine against *P.multocida* type A of sheep origin was prepared and compared with the whole cell vaccine, and it produced higher immune responses on using montanoide oil adjuvant. Economic loss of more than Rs 225 million was estimated due to haemorrhagic septicaemia in cattle and buffalo.

Gastrointestinal parasitism In Rajasthan software 'FROGIN' was evaluated for forecasting of gastrointestinal (GI) nematodosis in semi-arid and arid regions. It gives results as predicted faecal egg counts (FEC) on start of month, intensity of FEC for next 60 day and pasture larval burden for that month. The Garole sheep was not found completely refractory to infection of *Haemonchus contortus*. *Haemonchus, Bunostomum, Nematodirus* and *Oesophagostomum* spp. were found in all the zones in Sikkim. 170 kDa polypeptide of larval antigen of *H. contortus* was recognized by 4 day sera (prepatent sera) of sheep in western blotting. Zymogram studies revealed that 120 and 170 kDa polypeptide belonged to metalloproteases based upon protease inhibitor studies. In the ES product, cysteine protease and GST (glutathione-S-tranferase) were identified, which are of immunodiagnostic and immunoprophylectic value. In *H.contortus* ES antigen 30-32 kDa polypeptide showed protease activity, which was inhibited by E–64 confirming it to be a cysteine protease. GST was confirmed in western blotting utilizing anti-GST antibody.

ES antigen was better than gut antigen of *Ascaris suum*. Dipstic ELISA was comparatively found more efficient than plate ELISA. No correlation could be established between worm burden and antibody titre in naturally infected sera of pig with *A.suum*. Immunodominant polypeptide in *Bunostomum* and *Oesophagostomum* spp. was identified, and a diagnostic kit for serodiagnosis was developed and revalidated. Allele specific PCR was applied to field population of larvae for detection of benzimidazole resistance and was compared to FECRT and EHA. Frequency of BZrr (homozygous BZ-resistant) larvae in population ranged from 73.39 to 100% in Northern Rajasthan. Effect of *Fec* B gene on resistance to GIN was conducted in sheep naturally infected with GINs.

Lower incidence was observed in Garole sheep. H-11 and H-gal-GP polypeptides of *H. contortus* are being utilized for immunoprophylaxis and studies on H-gal-GP were completed. In H-gal- GP of *H. contortus* MEP-2 fragment showed 94% homology to other international strains. Equines Nation-wide active equine disease surveillance, sero-survey was conducted at Rajasthan, Haryana, Punjab, Uttar Pradesh, Madhya Pradesh, and Jammu and Kashmir. Antibodies to EHV-1 were detected in 7.1% samples, *Babesia equi* in 24.3% sera tested, and Japanese encephalitis in 5.5% serum samples tested.

None of the serum samples tested was positive for equine infectious anemia, African horse sickness, equine influenza and *Salmonella* Abortus-equi. Outbreaks of glanders reported from Uttarakhand, Andhra Pradesh, Himachal Pradesh and Haryana were investigated and diagnosed, and etiological agent *Burkholderia mallei* was islolated. Comparative sero-prevalence of JE in different animal species (equine, cattle, buffalo, pigs) was done in different regions of Haryana and highest incidence was in buffaloes followed by pigs, horses and cattle. EHV-4 virus was isolated from 11 out of 138 samples using equine embryonic lung cells. These results were confirmed by sequencing of PCR products. The centre succeeded in *in vitro* cultivation of bloodstream forms of *Trypanosoma evansi*. The characterization of *T. evansi* antigen by SDS-PAGE of sonicated antigen revealed five major polypeptides in the molecular weight range of 41-81 kDa and proteins of 35-41 kDa exhibited proteolytic activity. Serum neutralization test (SNT) and haemagglutination inhibition (HAI) were standardized for specific differentiation of two related arboviruses i.e., Japanese encephalitis (JE) and West Nile virus (WNV). On comparison sensitivity of HAI was 96.29% and specificity 100% in comparison to SNT. RT-PCR using Egene (291 bp) was also developed for diagnosis of JE in equines. An ELISA was developed for detection of *Babesia equi* specific antibodies. The sensitivity and specificity of the ELISA in comparison to commercial CI-ELISA was 94 and 96%, respectively. To study the polymorhphism of the MHC class II gene in Marwari horses, regions of MHC class-II (DRB-2a and 2b) gene fragments of 276 bp and 229 bp were amplified. Restriction analysis revealed that MHC-DRB2 (276 bp fragment) on digestion with *Hinf*I exhibits polymorphism in 48.39% genotypes. Microsatellite based parentage testing was done using 194 DNA samples collected from blood leukocytes of different horse breeds. Genotyping was performed by analysis of nine microsatellite and selected microsatellites were highly polymorphic as mean number of alleles ranged from 3.78 to 10.78. Total exclusionary power of both parents in all breeds was more than 0.9 and all the foals qualified the offspring-candidate parent compatibility.

Yak

A slide enzyme linked immunosorbent assay (SELISA) was standardized for the detection of *Babesia bigemina* antibodies in yak sera. Serological studies for detection of *B. bigemina* specific antibodies in yak from an organized farm and under field conditions revealed 44.16 and 56.10% seropositivity, respectively. It could be concluded that being more economical and technically simpler, SELISA could be used for seroprevalence studies on babesiosis in yak. Keratoconjunctivitis was noticed in yaks, and *Moraxella bovis* and *Neisseria* were recovered from the ocular swabs.

Serum samples from the affected animals were analyzed by viral neutralization test. ABELISA for the presence of bovine herpes virus - 1 (BHV-1) specific antibody. Nested PCR conducted using glycoprotein B and glycoprotein E specific primers (of BHV-1) revealed the presence of BHV- 1 in the ocular swab of the affected yaks. The serological and molecular analyses indicated the possible role of BHV-1 in severe forms of keratoconjunctivitis in yaks. Yak sera samples collected from different yak tracts of India were screened for the detection of BHV-1 specific antibody. The overall prevalence of BHV-1 specific antibody was alarmingly high (more than 40%) in yaks. Sex and location of different yak tracts did not have any influence over the IBR prevalence. However, the prevalence increased with the age of the animals, and was highest in yaks more than 3-year-old. Conjunctivitis and reproductive abnormalities were predominant symptoms among the seropositive yaks. The common ecological niche for feeding, watering and grazing with other domestic and wild animals is the possible avenue of infection in yaks.

Poultry

Marek's disease virus (MDV) circulating in PDP flocks was low virulent strain. MD incidence could be reduced more effectively with HVT (cell free) double dose or HVT+SB1 cell associated vaccines than single dose of HVT vaccine. Tumour samples could be safely stored in phenol-chloroformisoamyl alcohol for PCR. Leg weakness associated with osteomyelitis caused by *Staphylococcus* spp. was recorded in young broilers, while aspergillosis was observed in female line of Gramapriya.

Fisheries

Marine fish landings and catch structure: The marine fish landings of India during the year 2007 has been estimated as 2.88 million tonnes with an increase of about 1.7 lakh tonnes (6.5%) against the estimate of the previous year. The pelagic finfishes constituted 57%, demersal fishes 25%, crustaceans 14%, and molluscs 4% of the total landings. The sector-wise contributions were — mechanized landings 68%, motorized landings 28% and the atisanal landings 4%. Among the commercially important groups, the landings of oil sardine (26% increase over the previous year), penaeid prawns (13.4%), Indian mackerel (26%), croakers (41.9%) and other clupeids (55.6%) recorded substantial increase over their previous year's landings. The landings of non-penaeid prawns (18.6% decrease from the previous year), ribbon fishes (44%), Bombay duck (4.8%), thread fin breams (16.3%) and cuttle fish (27.2%) recorded marginal to substantial reduction from their previous year's landings. The estimates of region-wise production in the total production were—north-east region 13.2%, south-east region 22.6%, north-west region 29.3%, and south-west region 34.9%. Ring seine fishery for oil sardine along the northern coast of Tamil Nadu: Oil sardine (*Sardinella longiceps*), is the most important pelagic resource on the west coast of India, and its occurrence along the east coast was considered sporadic and rare. In July 2008, large shoals of oil sardine appeared in the near shore coastal waters of Devanampattinam, Cuddalore and Puducherry. This supported high catches in the ring seine units newly introduced fishing practice, from the near shore waters for a month. Inland fisheries Exotic fish species invasion in West Bengal wetlands – a cause of concern: During fish stock assessment, breeding populations of exotic tropical South American Sailfin catfish, *Pterygoplichthys disjunctivus* and *Pterygoplichthys pardalis,* were recorded in Gomokpota *beel* under East Kolkata Wetlands. Huge biomass of these species, approximately 20 metric tonnes, was caught in a single month. These fishes do not fetch remunerative price as food fish but occasionally find place in aquaria.

These species are prolonged breeders and voracious detritivores. A number of other exotic fish species, viz. *Barbonymus gonionotus* (*Puntius gonionotus*), *Pangasianodon hypophthalmus* (*Pangasius sutchi*), *Clarias gariepinnus, Oreochromis niloticus niloticus* and *Piaractus brachypomus* were also recorded in some other wetlands in the state. This has serious ecological and economic implications for the wetland fisheries in the state.

Culture Fisheries

Reservoir fisheries enhancement: Fisheries enhancement in reservoirs Dahod in Madhya Pradesh and Pahuj in Uttar Pradesh, was attempted through stocking of fish seed and improvement of institutional arrangements for fish catch and marketing. This resulted in improved fish production by over 60% in one year of experimental intervention. The catch/month, total fish catch and per month fishing days increased. The case studies will be helpful in formulating strategies for reservoir development in the Indo-Gangetic basin.

Freshwater Aquaculture

Labeo gonius in polyculture system: The compatibility of *Labeo gonius* with other major carps was studied through three combinations catla, silver carp, rohu and gonius; catla, silver carp, mrigal and gonius; and catla, silver carp, rohu and mrigal at combined density of 7,500 fingerlings/ha. Silver carp gave higher overall species survival, while catla showed the lowest level. Survival of rohu, mrigal and gonius, was intermediate and did not differ among them.

Coastal Aquaculture

Low fish meal feed for shrimp: A low fish meal shrimp feed was developed by replacing fish meal and other marine protein sources with plant protein sources. Shrimp production after four months of feeding on the low fish meal feed was 1,308 kg shrimp/ha with a feed conversion ratio (FCR) of 1.31:1. This feed can be successfully used for culturing tiger shrimp at low cost of production. WSSV risks to shrimp farming due to increased culture of crabs: Crabs are known carriers of white spot syndrome virus (WSSV), hence to address apprehensions of shrimp farmers to crab culture, studies were carried out to assess WSSV risks to shrimp farming due to enhanced culture of crabs.

The prevalence of WSSV in crustaceans in different geographical regions was estimated based on samples from Andhra Pradesh, Tamil Nadu, Maharashtra and West Bengal. The prevalence levels in crabs used for crab fattening or those found in wild crabs indicate that they do not pose any additional WSSV risks.

International Accounting Standard 41

Agriculture

The objective of this Standard is to prescribe the accounting treatment and disclosures related to agricultural activity.

This Standard shall be applied to account for the following when they relate to agricultural activity:

(a) biological assets;

(b) agricultural produce at the point of harvest; and

(c) government grants covered by paragraphs 34–35.

This Standard does not apply to:

(a) land related to agricultural activity; and

(b) intangible assets related to agricultural activity.

This Standard is applied to agricultural produce, which is the harvested product of the entity's biological assets, only at the point of harvest. Thereafter, IAS 2 *Inventories* or another applicable Standard is applied.

Accordingly, this Standard does not deal with the processing of agricultural produce after harvest; for example, the processing of grapes into wine by a vintner who has grown the grapes. While such processing may be a logical and natural extension of agricultural activity, and the events taking place may bear some similarity to biological transformation, such processing is not included within the definition of agricultural activity in this Standard.

The table below provides examples of biological assets, agricultural produce, and products that are the result of processing after harvest:

Biological assets	*Agricultural produce*	*Products that are the result of processing after harvest*
Sheep	Wool	Yarn, carpet
Plants	Cotton	Thread, clothing
Dairy cattle	Milk	Cheese
Pigs	Carcass	Sausages, cured hams
Bushes	Leaf	Tea, cured tobacco

Agriculture-related Definitions

The following terms are used in this Standard with the meanings specified:

- *Agricultural activity* is the management by an entity of the biological transformation and harvest of biological assets for sale or for conversion into agricultural produce or into additional biological assets.

- *Agricultural produce* is the harvested product of the entity's biological assets. A *biological asset* is a living animal or plant.
- *Biological transformation* comprises the processes of growth, degeneration, production, and procreation that cause qualitative or quantitative changes in a biological asset.
- A *group of biological assets* is an aggregation of similar living animals or plants.
- *Harvest* is the detachment of produce from a biological asset or the cessation of a biological asset's life processes.
- *Costs to sell* are the incremental costs directly attributable to the disposal of an asset, excluding finance costs and income taxes.

Agricultural activity covers a diverse range of activities; for example, raising livestock, forestry, annual or perennial cropping, cultivating orchards and plantations, floriculture, and aquaculture (including fish farming). Certain common features exist within this diversity:

(a) *Capability to change.* Living animals and plants are capable of biological transformation;

(b) *Management of change.* Management facilitates biological transformation by enhancing, or at least stabilising, conditions necessary for the process to take place (for example, nutrient levels, moisture, temperature, fertility, and light). Such management distinguishes agricultural activity from other activities. For example, harvesting from unmanaged sources (such as ocean fishing and deforestation) is not agricultural activity; and

(c) *Measurement of change.* The change in quality (for example, genetic merit, density, ripeness, fat cover, protein content, and fibre strength) or quantity (for example, progeny, weight, cubic metres, fibre length or diameter, and number of buds) brought about by biological transformation or harvest is measured and monitored as a routine management function.

Biological transformation results in the following types of outcomes:

(a) asset changes through

  (i) growth (an increase in quantity or improvement in quality of an animal or plant),

   (ii) degeneration (a decrease in the quantity or deterioration in quality of an animal or plant), or
   (iii) procreation (creation of additional living animals or plants); or

(b) production of agricultural produce such as latex, tea leaf, wool, and milk.

General Definitions

The following terms are used in this Standard with the meanings specified:

(a) the items traded within the market are homogeneous;

(b) willing buyers and sellers can normally be found at any time; and

(c) prices are available to the public.

- *Carrying amount* is the amount at which an asset is recognised in the statement of financial position.
- *Fair value* is the amount for which an asset could be exchanged, or a liability settled, between knowledgeable, willing parties in an arm's length transaction.
- *Government grants* are as defined in IAS 20 *Accounting for Government Grants and Disclosure of Government Assistance.*

The fair value of an asset is based on its present location and condition. As a result, for example, the fair value of cattle at a farm is the price for the cattle in the relevant market less the transport and other costs of getting the cattle to that market.

Recognition and Measurement

An entity shall recognise a biological asset or agricultural produce when, and only when:

(a) the entity controls the asset as a result of past events;

(b) it is probable that future economic benefits associated with the asset will flow to the entity; and

(c) the fair value or cost of the asset can be measured reliably.

In agricultural activity, control may be evidenced by, for example, legal ownership of cattle and the branding or otherwise marking of the cattle on acquisition, birth, or weaning. The future benefits are normally assessed by measuring the significant physical attributes.

A biological asset shall be measured on initial recognition and at the end of each reporting period at its fair value less costs to sell, except for the case described in paragraph 30 where the fair value cannot be measured reliably.

Agricultural produce harvested from an entity's biological assets shall be measured at its fair value less costs to sell at the point of harvest. Such measurement is the cost at that date when applying IAS 2 *Inventories* or another applicable Standard.

The determination of fair value for a biological asset or agricultural produce may be facilitated by grouping biological assets or agricultural produce according to significant attributes; for example, by age or quality.

An entity selects the attributes corresponding to the attributes used in the market as a basis for pricing. Entities often enter into contracts to sell their biological assets or agricultural produce at a future date. Contract prices are not necessarily relevant in determining fair value, because fair value reflects the current market in which a willing buyer and seller would enter into a transaction. As a result, the fair value of a biological asset or agricultural produce is not adjusted because of the existence of a contract. In some cases, a contract for the sale of a biological asset or agricultural produce may be an onerous contract, as defined in IAS 37 *Provisions, Contingent Liabilities and Contingent Assets*. IAS 37 applies to onerous contracts.

If an active market exists for a biological asset or agricultural produce, in its present location and condition, the quoted price in that market is the appropriate basis for determining the fair value of that asset. If an entity has access to different active markets, the entity uses the most relevant one. For example, if an entity has access to two active markets, it would use the price existing in the market expected to be used.

If an active market does not exist, an entity uses one or more of the following, when available, in determining fair value:

(a) the most recent market transaction price, provided that there has not been a significant change in economic circumstances between the date of that transaction and the end of the reporting period;

(b) market prices for similar assets with adjustment to reflect differences; and

(c) sector benchmarks such as the value of an orchard expressed per export tray, bushel, or hectare, and the value of cattle expressed per kilogram of meat.

In some cases, the information sources listed in paragraph 18 may suggest different conclusions as to the fair value of a biological asset or agricultural produce. An entity considers the reasons for those differences, in order to arrive at the most reliable estimate of fair value within a relatively narrow range of reasonable estimates.

In some circumstances, market-determined prices or values may not be available for a biological asset in its present condition. In these circumstances, an entity uses the present value of expected net cash flows from the asset discounted at a current market-determined rate in determining fair value.

The objective of a calculation of the present value of expected net cash flows is to determine the fair value of a biological asset in its present location and condition. An entity considers this in determining an appropriate discount rate to be used and in estimating expected net cash flows. In determining the present value of expected net cash flows, an entity includes the net cash flows that market participants would expect the asset to generate in its most relevant market. An entity does not include any cash flows for financing the assets, taxation, or re-establishing biological assets after harvest (for example, the cost of replanting trees in a plantation forest after harvest). In agreeing an arm's length transaction price, knowledgeable, willing buyers and sellers consider the possibility of variations in cash flows. It follows that fair value reflects the possibility of such variations.

Accordingly, an entity incorporates expectations about possible variations in cash flows into either the expected cash flows, or the discount rate, or some combination of the two. In determining a discount rate, an entity uses assumptions consistent with those used in estimating the expected cash flows, to avoid the effect of some assumptions being double-counted or ignored.

Cost may sometimes approximate fair value, particularly when:

(a) little biological transformation has taken place since initial cost incurrence (for example, for fruit tree seedlings planted immediately prior to the end of a reporting period); or

(b) the impact of the biological transformation on price is not expected to be material (for example, for the initial growth in a 30-year pine plantation production cycle).

Biological assets are often physically attached to land (for example, trees in a plantation forest). There may be no separate market for biological assets that are attached to the land but an active market may exist for the combined assets, that is, for the biological assets, raw land, and land improvements, as a package.

An entity may use information regarding the combined assets to determine fair value for the biological assets. For example, the fair value of raw land and land improvements may be deducted from the fair value of the combined assets to arrive at the fair value of biological assets.

Gains and Losses

A gain or loss arising on initial recognition of a biological asset at fair value less costs to sell and from a change in fair value less costs to sell of a biological asset shall be included in profit or loss for the period in which it arises. A loss may arise on initial recognition of a biological asset, because costs to sell are deducted in determining fair value less costs to sell of a biological asset. A gain may arise on initial recognition of a biological asset, such as when a calf is born. A gain or loss arising on initial recognition of agricultural produce at fair value less costs to sell shall be included in profit or loss for the period in which it arises. A gain or loss may arise on initial recognition of agricultural produce as a result of harvesting.

Inability to Measure Fair Value Reliably

There is a presumption that fair value can be measured reliably for a biological asset. However, that presumption can be rebutted only on initial recognition for a biological asset for which market-determined prices or values are not available and for which alternative estimates of fair value are determined to be clearly unreliable. In such a case, that biological asset shall be measured at its cost less any accumulated depreciation and any accumulated impairment losses. Once the fair value of such a biological asset becomes reliably measurable, an entity shall measure it at its fair value less costs to sell. Once a non-current biological asset meets the criteria to be classified as held for sale (or is included in a disposal group that is classified as held for sale) in accordance with IFRS 5 *Non-current Assets Held for Sale and Discontinued Operations*, it is presumed that fair value can be measured reliably.

The presumption in paragraph 30 can be rebutted only on initial recognition. An entity that has previously measured a biological asset at its fair value less costs to sell continues to measure the biological asset at its fair value less costs to sell until disposal. In all cases, an entity measures agricultural produce at the point of harvest at its fair value less costs to sell. This Standard reflects the view that the fair value of agricultural produce at the point of harvest can always be measured reliably.

In determining cost, accumulated depreciation and accumulated impairment losses, an entity considers IAS 2 *Inventories*, IAS 16 *Property, Plant and Equipment* and IAS 36 *Impairment of Assets*.

Government Grants

An unconditional government grant related to a biological asset measured at its fair value less costs to sell shall be recognised in profit or loss when, and only when, the government grant becomes receivable.

If a government grant related to a biological asset measured at its fair value less costs to sell is conditional, including when a government grant requires an entity not to engage in specified agricultural activity, an entity shall recognise the government grant in profit or loss when, and only when, the conditions attaching to the government grant are met. Terms and conditions of government grants vary. For example, a grant may require an entity to farm in a particular location for five years and require the entity to return all of the grant if it farms for a period shorter than five years. In this case, the grant is not recognised in profit or loss until the five years have passed.

However, if the terms of the grant allow part of it to be retained according to the time that has elapsed, the entity recognises that part in profit or loss as time passes.

If a government grant relates to a biological asset measured at its cost less any accumulated depreciation and any accumulated impairment losses, IAS 20 *Accounting for Government Grants and Disclosure of Government Assistance* is applied.

This Standard requires a different treatment from IAS 20, if a government grant relates to a biological asset measured at its fair value less costs to sell or a government grant requires an entity not to engage in specified agricultural activity. IAS 20 is applied only to a government grant related to a biological asset measured at its cost less any accumulated depreciation and any accumulated impairment losses.

Disclosure

General

An entity shall disclose the aggregate gain or loss arising during the current period on initial recognition of biological assets and agricultural produce and from the change in fair value less costs to sell of biological assets.

An entity shall provide a description of each group of biological assets. The disclosure required by paragraph 41 may take the form of a narrative or quantified description. An entity is encouraged to provide a quantified description of each group of biological assets, distinguishing between consumable and bearer biological assets or between mature and immature biological assets, as appropriate. For example, an entity may disclose the carrying amounts of consumable biological assets and bearer biological assets by group. An entity may further divide those carrying amounts between mature and immature assets. These distinctions provide information that may be helpful in assessing the timing of future cash flows. An entity discloses the basis for making any such distinctions. Consumable biological assets are those that are to be harvested as agricultural produce or sold as biological assets. Examples of consumable biological assets are livestock intended for the production of meat, livestock held for sale, fish in farms, crops such as maize and wheat, and trees being grown for lumber. Bearer biological assets are those other than consumable biological assets; for example, livestock from which milk is produced, grape vines, fruit trees, and trees from which firewood is harvested while the tree remains. Bearer biological assets are not agricultural produce but, rather, are self-regenerating.

Biological assets may be classified either as mature biological assets or immature biological assets. Mature biological assets are those that have attained harvestable specifications (for consumable biological assets) or are able to sustain regular harvests (for bearer biological assets).

If not disclosed elsewhere in information published with the financial statements, an entity shall describe:

(a) the nature of its activities involving each group of biological assets; and

(b) non-financial measures or estimates of the physical quantities of:

  (i) each group of the entity's biological assets at the end of the period; and

  (ii) output of agricultural produce during the period.

An entity shall disclose the methods and significant assumptions applied in determining the fair value of each group of agricultural produce at the point of harvest and each group of biological assets. An entity shall disclose the fair value less costs to sell of agricultural produce harvested during the period, determined at the point of harvest.

An entity shall disclose:

(a) the existence and carrying amounts of biological assets whose title is restricted, and the carrying amounts of biological assets pledged as security for liabilities;

(b) the amount of commitments for the development or acquisition of biological assets; and

(c) financial risk management strategies related to agricultural activity.

An entity shall present a reconciliation of changes in the carrying amount of biological assets between the beginning and the end of the current period. The reconciliation shall include:

(a) the gain or loss arising from changes in fair value less costs to sell;

(b) increases due to purchases;

(c) decreases attributable to sales and biological assets classified as held for sale (or included in a disposal group that is classified as held for sale) in accordance with IFRS 5;

(d) decreases due to harvest;

(e) increases resulting from business combinations;

(f) net exchange differences arising on the translation of financial statements into a different presentation currency, and on the translation of a foreign operation into the presentation currency of the reporting entity; and

(g) other changes.

The fair value less costs to sell of a biological asset can change due to both physical changes and price changes in the market. Separate disclosure of physical and price changes is useful in appraising current period performance and future prospects, particularly when there is a production cycle of more than one year. In such cases, an entity is encouraged to disclose, by group or otherwise, the amount of change in fair value less costs to sell included in profit or loss due to physical changes and due to price changes. This information is generally less useful when the production cycle is less than one year (for example, when raising chickens or growing cereal crops). Biological transformation results in a number of types of physical change—growth, degeneration, production, and procreation, each of which is observable and measurable. Each of those physical changes has a direct relationship to future economic benefits. A change in fair value of a biological asset due to harvesting is also a physical change.

Agricultural activity is often exposed to climatic, disease and other natural risks. If an event occurs that gives rise to a material item of income or expense, the nature and amount of that item are disclosed in accordance with IAS 1 *Presentation of Financial Statements*. Examples of such an event include an outbreak of a virulent disease, a flood, a severe drought or frost, and a plague of insects.

Additional disclosures for biological assets where fair value cannot be measured reliably. If an entity measures biological assets at their cost less any accumulated depreciation and any accumulated impairment losses. At the end of the period, the entity shall disclose for such biological assets:

(a) a description of the biological assets;

(b) an explanation of why fair value cannot be measured reliably;

(c) if possible, the range of estimates within which fair value is highly likely to lie;

(d) the depreciation method used;

(e) the useful lives or the depreciation rates used; and

(f) the gross carrying amount and the accumulated depreciation (aggregated with accumulated impairment losses) at the beginning and end of the period.

If during the current period, an entity measures biological assets at their cost less any accumulated depreciation and any accumulated impairment losses, an entity shall disclose any gain or loss recognised on disposal of such biological assets and the reconciliation required by paragraph 50 shall disclose amounts related to such biological assets separately. In addition, the reconciliation shall include the following amounts included in profit or loss related to those biological assets:

(a) impairment losses;

(b) reversals of impairment losses; and

(c) depreciation.

If the fair value of biological assets previously measured at their cost less any accumulated depreciation and any accumulated impairment losses becomes reliably measurable during the current period, an entity shall disclose for those biological assets:

(a) a description of the biological assets;

(b) an explanation of why fair value has become reliably measurable; and

(c) the effect of the change.

Government Grants

An entity shall disclose the following related to agricultural activity covered by this Standard:

(a) the nature and extent of government grants recognised in the financial statements;

(b) unfulfilled conditions and other contingencies attaching to government grants; and

(c) significant decreases expected in the level of government grants.

Need for Accounting Standards for Agriculture and Livestock

According to the NAFSCOB (National Federation of State Co-operative Banks Ltd) report there are 30 SCBs (State Co-operative Banks), 962 branches, 370 DCCBs (Districts Central Co-operative Banks and 1,06,384 primary societies in India by 31/3/2006.

The Government of India established NABARD to finance the agricultural sector through state and District Co-operative Banks. All the DCCBs and PACS (Primary Agricultural Credit Societies) are lending loans and advances for the development of agricultural sector and uplifting of weaker section.

The DCCBs and PACBs in India provide loans and advances for different agricultural schemes. These schemes are formulated and approved under the supervision and control of NABARD. The high yielding cultivable crops are identified for different regions and financial assistances are offered under Short Term and Medium Term (ST & MT), Schematic lending, ST & MT Agricultural loans and SAOs (Seasonal and Agricultural Operations) Financial assistance is provided for marketing, storage, purchase of seeds, fertilizers and pesticides and for all the agricultural activities.

The commercial crops are also identified and loans and advances are provided for production, marketing and for other related needs under separate schemes and plans. In the books of accounts of any DCCB or PACS we can find plenty of loans and advances provided for cattle purchase, poultry, goat farm etc. In the accounting parlance these are brought under the category of 'Livestock'. As per schedule VI of the Companies Act the livestock is included in fixed assets. Normally fixed assets are brought under the purview of depreciation as per the generally acceptable accounting principles and standard practices. It is obligatory as per legal provisions of certain Acts. The stock of inventories is categorized as current asset and valued according to the standard practices followed by the respective organizations.

According to AS 9 - Revenue Recognition concept, the minimum amount of sale income earned from agriculture can be reliably measured even at the time of completion of production. This may be due to support price set by the Government or immediate marketability.

The price is pre-determined before marketing it. In such case the revenue may be recognized as soon as the production or harvesting is completed at the pre-determined price and that will be the selling price. There are so many crops for which no adequate support price is being provided and immediate convertibility into cash is also not possible. We could find a dispute in existing support price or the price determined every year by the Govt.

The IAS-41 (International Accounting Standard) has been introduced for Agriculture. The General Clarification (GC) - 12/2002, issued by the Accounting Standard Board of the ICAI on applicability of accounting standards indicated that the Accounting standards would apply to all the activities of the co-operative societies including those which are not commercial, Industrial and or business in nature.

The Auditors should examine the books of accounts of them in accordance with Accounting Standards and deviations must be disclosed. According to Accounting Standard 2 revised (AS-2 revised) the stock of inventories that is one of the current assets must be valued at 'Net Realizable Value'. The AS-2 is not applicable to producer's inventories of livestock, agricultural and forest products, mineral oils, ores and gases.

According to AS-2 and IAS-2 (International Accounting Standards) they are measured at net realizable value in accordance with well-established practices in those concerns. It is obligatory to follow similar accounting policies consistently in all accounting periods. Any change must be disclosed in the financial statement. AS-10 deals with Fixed Assets. All the depreciable assets must be depreciated as per the standard practices. The depreciation is calculated under SLM (Straight Line Method), WDV (Written down Value method), unit of production method or any other standard method permitted to follow. Now the questions raised on stock of agricultural produce and live stock are :

1. What are the procedures for recording such items?
2. What are the methods of valuing such stock?
3. If the current practices followed are approved which are the Standard practices to be considered prudential?

When these questions are raised to a group of practicing Charted Accountants, they said it is an unimportant area and hence they accept the current method of valuation followed by their client for agricultural produces or livestock. Since it is neglected as an unimportant area, proper emphasis was not given so far. But, India is an agricultural based country and more than 90% of village population has livestock. Most of the farmers and villagers earn their livelihood from livestock and income from crops. They take crop loan and schematic loans from DCCBs and PACSs available in their respective jurisdiction.

Even the authorities concerned accept that they do not follow any standard practice for valuing the yield of livestock. The breeding animals loan and all kinds of loan granted for livestock are highly hypothetical. It seems that it is granted on trial and error basis.

In practice the DCCB and PACS ask the party to produce a "Veterinary Doctor Certificate" about the health condition of the cattle or the livestock to approve the loan. Many of their methods are substandard and unrealistic. The agricultural loans and crops loans are provided based on the agricultural schemes and plans that are prepared on unrealistic assumptions on the value of yield and duration of the yield. So in order to develop standard practices, the Indian Accounting Standard Board must come forward to formulate an Accounting Standard in the form of guidelines or provisions.

While formulating it, cautions must be taken to prevent contradictions with other Accounting Standards, Concepts and Conventions, especially with revenue recognition concept and matching concept. It is expected that this will give a convincing solution for the following questions.

1. What are the Principles to be followed in recording Agriculture produce and livestock in books of accounts?
2. What are the methods of valuing the yield before and after the harvest of commercial and non-commercial crops?
3. What are the situations in which livestock can be considered as fixed assets?
4. What are the situations in which the livestock can be considered as current Assets?
5. How to value the Livestock?

This would help to develop standard and productive practices in granting loan to develop rural population in the long run. The lending policies and collection of overdue would not thrust any burden or risk. The people will use the rural credit wisely only on productive schemes.

The unscrupulous persons could be prevented from misusing such facilities available in co-operative banks. This would facilitate to reach the schemes and plans to the right people. The people will feel their moral responsibility of repaying the loan and the NPA (Non- Performing Assets) in the DCCBs and PACS in India could be brought under control.

Industrial Livestock Production and Global Health Risks

Recent emergence of contagious human diseases from animal populations, such as Nipah virus infection in 1999, SARS in 2002 and the current epidemic of highly pathogenic avian influenza (HPAI) caused by H5N1, from which nearly 200 people have died since 2004, have heightened public awareness of possible linkages between wild animal populations, livestock production and global public health.

Because of human and livestock population growth, changes in livestock production, the emergence of worldwide agro-food networks, wild animal trade, and significant changes in personal mobility, human populations increasingly share a global commons of disease risk, among themselves and with other animal species. It is, therefore, not surprising that three out of four emerging pathogens affecting humans over the past ten years have originated from animals or animal products (Taylor *et al.*, 2001). HPAI offers a recent example of how a new viral challenge has possibly emerged from wildlife, by first adapting to domestic livestock and then circulating within these populations with the risk of acquiring the ability to infect humans and of sustained human-to-human transmissibility.

The case of HPAI also highlights how conditions of animal husbandry and the livestock supply chain can influence health risks for human populations worldwide. While individual countries have taken steps to contain outbreaks and to dissemination of HPAI, pathogens can move by unregulated and unrecognized pathways, such as on airborne dust, in animal wastes utilized in agriculture, in ballast water on ships, and in migrating wild animals. It is imperative to understand the formal and informal networks of exchange in order to develop evidence-based policies to anticipate and prevent emergence of novel zoonoses. To elucidate the linkage between livestock production and the global commons of disease risk, we draw upon recent experiences provided by different influenza A virus (IAV) incursions into domestic livestock populations, the most notable one being the ongoing HPAI H5N1 epidemic that originated in Asia, which now also affects Africa and with outbreaks having been recorded in the Near East and in Europe.

This paper reviews changes in food animal production that have occurred over the past decade, and then summarizes the current understanding of the emergence of distinct influenza viruses.

Direct and indirect evidence, drawn primarily from HPAI outbreaks in areas of high poultry density, is presented on aspects of biosecurity and biocontainment that are relevant to the transmission of influenza viruses in industrial poultry production systems. Incentives associated with the management of animal and public health risks are discussed based on the evidence presented.

Changes in Food Animal Production

As countries have become more affluent, demand for livestock-derived food has substantially increased, leading to a major transformation of global animal food production. The linkages between sub-sectors of the animal industry, such as feed manufacturers, breeding companies, livestock keepers and processors, as well as production practices have changed significantly over the past decades, with potentially serious consequences for disease risks.

These changes include significant increases in livestock populations and densities, concentrated industrial food animal production, using fewer but more productive livestock breeds and lines, with, in the case of poultry and pigs, hybrid animals providing the end product, specialization in and vertical integration of stages of production (e.g. breeding, raising, finishing), and major changes in the design of animal housing facilities.

Industrial food animal production involves high throughput animal husbandry, in which thousands of animals of similar genotypes are raised for one purpose (such as pigs, layer hens, broiler chickens, ducks, turkeys) with rapid population turnover at one site under highly controlled conditions, often in confined housing, with nutrient dense, industrial feeds replacing access to forage crops. In the US, these facilities are known as animal feeding operations (AFOs). Concentrated animal feeding operations (CAFOs) are a type of AFO, which have a regulatory definition in the US as facilities that have animals stabled or confined for at least 45 days out of any 12 month period and holding at least 1,000 animal units (AUs) (1 AU = 1000 pounds body weight).

Globally, pig and poultry production are the fastest growing and industrializing livestock subsectors with annual production growth rates of 2.6 and 3.7 percent over the past decade. As a consequence, in the industrialized countries, the vast majority of chickens and turkeys are now produced in houses in which between 15,000 and 50,000 birds are confined throughout their lifespan. Increasingly, pigs and cattle are also raised under similar conditions of confinement and high density. The trend towards industrialization of livestock production can also be observed in developing countries, where traditional systems are being replaced by intensive units at a rate of 4.3 percent of animal holding units per year, with much of that increase occurring in Asia, South America and North Africa (CAST, 1999). In developing countries a large proportion of industrial units are sited in or close to human population centres. Over the same time, the human population has grown by almost 700 million people, again, with much of this growth occurring in the developing world and in particular affecting urban populations. Industrialization of food animal production has led to major increases in livestock productivity, which are to a large extent the result of genetic progress and the development of diets tailored to specific stages of production. For poultry and pigs, industrial production is organized in stages which separate primary breeders, multipliers and producers (often contract farmers). A small number of globally operating companies form the apex of the breeding pyramid. Different production stages are often undertaken at different sites, leading to significant movement of live animals, at times across national borders.

In 2005, for example, more than 25 million live pigs, i.e. more than 2 million pigs per month, were traded internationally (FAOSTAT). In the US, there is a huge movement of unfinished animals, for example feeder pigs from the Carolinas to the Cornbelt. In 2001, 27 percent of pigs in the US were moved from one state to another (or more) (Hennessy, 2004). Investigations in relation to the recent HPAI outbreak in the UK revealed that links in poultry production within one enterprise between facilities located in the UK and in Hungary involved movement of hatching eggs, birds and poultry products four times before the final product reached retail (Lucas, 2007). Animal slaughter operations have also become concentrated, leading to larger average distances for transport to slaughter (Burrell, 2002; MLC, 2001).

The consolidation of poultry and pig production for reasons of competitive advantage has also affected the geography of food animal populations. Over the past 60 years, the geographic distribution of both pig and poultry production in the US, for example, has become more clustered, with poultry production now being highly concentrated in the southeastern states and pig production concentrated in some of these same states, as well as in the Midwest. Similar trends have occurred worldwide with pig and poultry populations increasingly concentrated in particular locations which are often geographically coincident.

Industrial pig and poultry production with its geographic intensity and being coincident for the two species, and with the regular movement of animals between production stages provides significant opportunities for interactions between large populations of confined poultry and/or pigs and thus has potential consequences for the development and transmission of some zoonotic disease agents. The proximity of thousands of confined animals increases the likelihood of transfer of pathogens within and between these populations, with consequent impacts on rates of pathogen evolution. Furthermore, animals held in confinement produce large amounts of waste, which need to be disposed of.

Much of this waste, which may contain large quantities of pathogens, is disposed of on land, posing an infection risk for wild mammals or avians. Poultry house waste is also utilized in aquaculture, a form of food animal production, which results in the creation of artificial wetlands and thereby increases direct opportunities for contact with wild avians.

Emergence of Novel Influenza A Viruses

Evidence suggests that non-domesticated aquatic birds are the primary reservoir of influenza A viruses (IAVs) and probably all IAVs of mammals have ancestral links to avian lineages (Webby and Webster, 2001). An important feature of IAVs is their capacity to undergo molecular transformation and to adapt to new host populations1 and thereby acquire the potential to cause major disease outbreaks.

Another criterion for classifying IAVs is their pathogenicity, i.e. their ability to cause disease in experimentally inoculated chickens. Thus virulent viruses leading to severe disease and high levels of mortality are classified as highly pathogenic AI (HPAI) while viruses causing much milder disease (primarily mild respiratory disease, depression and reduction in egg production in laying birds) are classified as low pathogenicity AI (LPAI).

To date, only IAVs of the H5 and H7 subtype have been shown to cause HPAI, but not all H5 and N7 IAVs are highly pathogenic. Current evidence suggests that HPAI viruses are not endemic in wild bird populations and only arise as a result of molecular changes from low pathogenicity IAVs (LPAI) that occur after introduction into domestic poultry from wild birds (Capua and Alexander, 2004), although one outbreak of HPAI H5N3 that could not be linked to domestic poultry has been recorded in terns in South Africa in 1961.

Evidence suggests that introduction of LPAI viruses into domestic poultry populations usually occurs as a result of direct or indirect contact with wild waterfowl or domestic ducks. Various incursions of LPAI virus into domestic poultry have been reported over the past decade, mostly in North America and Europe, but also in Mexico, Chile and Pakistan, details of which have been compiled by Capua and Alexander (2004). In Mexico, 1994, a LPAI H5N2 virus mutated into a HPAI virus and spread to Guatemala in 2000 and to El Salvador in 2001, presumably via trade (ibid). LPAI H5N2 is now established in domestic chicken populations in Central America. In both, the 2003 H7N7 HPAI epidemic in the Netherlands (Stegeman *et al.*, 2004) and the 2004 H7N3 HPAI epidemic in British Columbia, Canada (Power, 2005), it appears that LPAI infections in poultry preceded the emergence of HPAI in different poultry houses on the same commercial farms. In Italy, the 1999/2000 H7N1 HPAI epidemic was preceded by 199 reported outbreaks of LPAI H7N1 in the same region. On the other hand, between the mid-1990s and 2004 H7N2 LPAI virus appears to have been endemic in parts of the US, linked to live poultry markets, without conversion to an HPAI virus.

Overall, over the past 10 years reports of HPAI have increased dramatically with ten distinct minor or major epidemics reported worldwide since 1997 while 14 outbreaks have been recorded over the preceding 40 years. It is however important to acknowledge the limits in detection and reporting of HPAI and LPAI outbreaks when attempting to interpret this apparent increase in disease incidence.

In addition to the apparent increase in outbreaks caused by H5 and H7 LPAI and HPAI viruses, H9N2 LPAI virus has also spread through domestic poultry populations in the 1990s and become endemic in commercial poultry in China, the Middle East and elsewhere. H9N3 outbreaks are said to have occurred in China in 1994 (Yingjie, 1998, quoted in Capua and Alexander, 2004).

Pigs may potentially assume an important role in the emergence of novel IAVs as they can be infected by either avian or human viruses (Kida *et al.*, 1994; Schulz *et al.*, 1991). Gilchrist *et al.* (2007) note the proximity of concentrated poultry and swine operations as a source of disease risk from IAVs although so far there have only been reports of AI viruses from poultry in pigs and not vice-versa. Classical H1N1 swine influenza viruses are very similar to the virus implicated in the 1918 – 1919 human influenza pandemic and circulate predominantly in the US and Asia.

H3N2 viruses of human origin have been isolated from pigs in Europe and the Americas shortly after their emergence in humans (Webby and Webster, 2001) and are now endemic in pigs in southern China (Peiris *et al.*, 2001), where they co-circulate with H9N2 viruses with the potential of reassortment with H5N1.2 In the USA, outbreaks of respiratory disease in swine herds have been caused by IAVs which arose from reassortment of human H3N2, classic swine H1N1 and avian viral genes (Zhou *et al.*, 1999) Thus, it appears that IAVs are now fairly widespread in commercial poultry (and to a lesser extent pig) populations and that highly pathogenic strains are emerging or being detected with growing frequency. This may be due to the substantial increase in poultry numbers (25 percent) over the last decade and/or to changes in poultry production, as well as to enhanced diagnostic capacities and better disease reporting and/or to changes in poultry production (Capua and Alexander, 2006). Although the specific role of CAFOs in the emergence of novel diseases is not well understood, the US Council for Agriculture, Science and Technology (CAST, 2005) warns that a major consequence of modern industrial livestock production systems is that they potentially allow the rapid selection and amplification of pathogens.

Biosecurity and Disease Transmission in Industrial Systems

Biosecurity has been broadly defined as any practice or system that prevents the spread of infectious agents from infected to susceptible animals, or prevents the introduction of infected animals into a herd, region, or country in which the infection has not yet occurred (Radostits, 2001). More specifically, farm biosecurity combines 'bioexclusion', i.e. measures for preventing a pathogen from being introduced to a herd/ flock, and 'biocontainment', which addresses events after introduction, i.e. the ability for a pathogen to spread amongst animal groups within a farm or of being released from the farm (Dargatz *et al.*, 2002). The majority of farms devote most resources to bioexclusion although large multiunit farms also invest in biocontainment to prevent spread to other units. Disease transmission between farms depends on the combination of individual bioexclusion practices and biocontainment measures. The importance of the latter is largely determined by the magnitude and direction of resource (and waste) flows across and between farm populations.

In the livestock sector, these flows can be complex because of specialization at different stages of animal production and processing and intricate formal and informal market chains. Large scale production at all stages is highly concentrated, with small numbers of large, intensive facilities and even fewer responsible enterprises. Larger facilities are often assumed to implement more advanced biosecurity measures, but the intensity of their operations also poses higher risks for infection and pathogen propagation. Over one cycle of 10,000 broilers for example, around 42 tons of feed and 100,000 l of water have to be supplied to the birds, and unless stringent measures are taken these remain potential routes of introduction, while around 20 tons of waste will be produced requiring disposal.

The design and operational requirements of large scale poultry and swine houses also result in compromises of biosecurity. Because confinement of thousands of animals requires controls to reduce heat and regulate humidity, poultry and swine houses are ventilated with high volume fans that result in considerable movement of materials into the external environment (Jones *et al.*, 2005) and dust emissions, visible (particles >10 microns) and invisible (particles <10 microns), from poultry barns housing thousands of birds can be substantial.

Measurement of aerosol emissions from a broiler operation revealed a million fold elevated concentration of aerosolized invisible dust near a poultry barn fan as compared to outdoor air in a semi-rural area (Power, 2005). These particles have the potential to remain suspended in the air for up to several days, and, depending on prevailing winds, poultry barn dust could be found several kilometers from its source. Although little is known about the survival of IAVs on dust particles, high concentrations of infectious AI virus have been detected in air samples from an infected barn while low levels of virus were detected in one of nine samples some 800 m from an infected barn. However, it was not determined if this sample of virus was infectious (Power, 2005).

Other pathogens have been shown to readily move in and out of poultry and swine houses. Pathogen entry was demonstrated in a recent study of *Campylobacter*-free broiler flocks, housed in sanitized facilities, using standard biosecurity measures, and fed *Campylobacter*-free feed and water. Seven out of ten flocks became colonized with *Campylobacter* by the time of slaughter and two flocks were colonized by *Campylobacter* strains genetically indistinguishable from strains isolated from puddles outside of the facility prior to flock placement (Bull *et al.*, 2006).

Although the route of entry was not determined, this study clearly shows that some pathogens in the immediate environment of a poultry facility have a high chance to overcome standard bioexclusion measures. Contaminated air exiting the house via ventilation systems becomes a source of *Campylobacter* to the external environment and microbes may be carried some distance by wind and surface water transport. *Campylobacter* strains with identical DNA fingerprints to those colonizing broilers have been measured in air up to 30 m downwind of broiler facilities housing colonized flocks (Lee *et al.*, 2002).

Insects are another means for pathogen entry to and exit from poultry houses. Research carried out during an HPAI outbreak in Kyoto, Japan in 2004, found that flies caught in proximity to broiler facilities where the outbreak took place, carried the same strains of H5N1 influenza virus as found in chickens of an infected poultry farm (Sawabe *et al.*, 2004). A study in Denmark found that as many as 30,000 flies may enter a broiler facility during a single flock rotation in the summer months (Hald *et al.*, 2004). Evidence that bioexclusion measures of (at least some) large-scale industrial poultry operations are not impenetrable by IAVs is provided by reviewing the HPAI H5N1 outbreaks reported to OIE (www.oie.int).

HPAI H5N1 for instance has been reported to have caused outbreaks in largescale industrial poultry units with supposedly high biosecurity standards in South Korea (a 300,000 bird unit), Russia (two 200,000 bird units) and Nigeria (a 50,000 bird unit) in 2006, and in the UK (a 160,000 turkey unit) in 2007. Moreover, large(r) industrial-type flocks appear to be overrepresented in the list of HPAI H5N1 outbreaks reported to OIE vis-à-vis outbreaks in backyard/village flocks in relation to their respective shares of total national flocks.

Around 40 percent of the HPAI H5N1 outbreaks in domestic poultry reported to OIE between late 2005 and early 2007 occurred in poultry units of 10,000 birds or more (more than 25 percent occurred in units of more than 10,000 birds), while, even in many OECD countries, e.g. Germany, France, UK and Belgium, less than 10 percent of flocks consist of more than 10,000 birds. It is likely that some of this overrepresentation of large industrial-type flocks in reported outbreaks is due to ascertainment bias because outbreaks are more likely to be detected and reported in large-scale operations than in backyard systems3. Nevertheless, it demonstrates that, whatever the source of the virus (wild avians, backyard poultry, other commercial units), bioexclusion measures implemented by some large-scale industrial poultry units, including those in industrialized countries, may be insufficient to protect against H5N1 incursion when challenged.

This empirical evidence of sub-optimal biosecurity of a proportion of commercial operations is substantiated by direct observations and points to a need for greater oversight and or regulation of biosecurity of industrial poultry production. For instance, Power (2005) reports that more than three quarters of commercial broiler and table egg farms in Fraser Valley, Canada, had indicated in a survey that they did not provide disinfection footbaths nor required a change of clothes/coveralls by employees on entering their barns. In Maryland, US, Price *et al.* (2007) found that most poultry workers are provided little or no protective clothing or opportunities for personal hygiene or decontamination on site, and that they take their clothes home for washing. Once IAVs have entered industrial production facilities they can be transferred between operations by contaminated shipping containers and trucks.

Given that a gram of infected faeces can contain as many as ten billion infectious virus particles, a small amount of contaminated faecal material or litter adhering to boots, clothing or equipment may be sufficient to transmit virus from an infected to a susceptible flock (Power, 2005). Biocontainment of IAV, once poultry are infected thus poses a substantial challenge, even in countries with advanced animal-health services and depends on early detection of outbreaks and action before the virus has spread widely in infected premises.

HPAI Epidemics in Densely Populated Poultry Production Areas

The 1999-2000 H7N1 epidemic in northern Italy, the 2003 H7N7 epidemic in the Netherlands and the 2004 H7N3 epidemic in Fraser Valley (British Columbia, Canada) highlight the difficulties faced by animal health authorities when HPAI infects flocks in densely populated poultry production areas (DPPAs).

In all three epidemics animal health authorities noted the high density of poultry farms with frequent contact between farms by trucks and surprisingly low levels of biosecurity practiced by some operators as having been associated with the considerable spread of virus (Capua *et al.*, 2002; Stegeman *et al.*, 2004; Power 2005). Retrospective analysis of between-flock transmission in two distinct outbreak areas in the case of the Netherlands and of an H7N3 LPAI epidemic in Italy in 2003-2004 estimated reproduction ratios (R*h*, the average number of secondary infections caused by 1 infectious flock) of 6.5, 3.1 and 2.9 respectively (Stegeman *et al.*, 2004; Capua and Marangon, in press) prior to the implementation of control measures.

This clearly indicates that standard bioexclusion and biocontainment measures in a number of the predominantly industrial flocks were insufficient to prevent disease spread, and that disease detection or reporting was delayed. For caged layers, for example, Capua and Alexander (2006b) cite a 'flock incubation period' of up to 18 days, which, in areas with intense betweenfarm traffic, provides sufficient time extensive movement of the virus.

Control of the three epidemics was only achieved through massive depopulation of commercial and backyard/hobby flocks (vaccination was not applied). Retrospective analysis of the Dutch outbreak also revealed that between-flock transmission continued even after the implementation of strict movement controls in the affected areas. The authors conclude that containment of the epidemic was more likely to be the result of the depletion of susceptible flocks by depopulation than the reduction of the transmission rate through biocontainment measures (Stegeman *et al.*, 2004).

The lower probability of infection of backyard/hobby flocks compared to that of industrial flocks in the Dutch and Canadian epidemics, in which samples were collected from backyard/hobby flocks in the vicinity of infected industrial flocks, is consistent with findings from the HPAI epidemic in Thailand in 2004 and the 2002 outbreak of Newcastle disease in Denmark (Otte *et al.*, 2007) and suggests that commercial transactions are an important route for disease transmission between industrial farms.

The Animal: Human Interface

Over the past 100 years, the sudden emergence of antigenetically different strains of IAVs transmissible among humans leading to human influenza pandemics has occurred in 1918 (H1N1, Spanish flu), 1957 (H2N2, Asian flu), and 1968 (H3N2, Hong Kong flu) (Webby and Webster, 2001). Studies on these pandemic viruses have shown that all three contained an avian component (Capua and Alexander, 2006b).

A number of studies demonstrate that IAVs from animals can move across the animal:human interface in the context of food animal production and processing, and therefore, livestock keepers and people otherwise in close contact with live animals are the most likely group to act as 'bridge' for IAVs between livestock and human communities at large, should an IAV acquire the capability of sustained human-to-human transmission.

Myers *et al.* (2006), for example, report that swine farmers had higher titres of H1N1 and H1N2 antibodies and greatly elevated risks of seropositivity to these two influenza A viruses (35.3 and 13.8 odds ratios respectively), as compared to community referents. A comprehensive study of the outcome of exposure to H5N1 has been conducted in Hong Kong during the 1997 outbreak (Bridges *et al.*, 2002). In this study, 1,525 poultry workers and 293 government workers involved in culling activities and disease investigations were assessed for risk factors for seropositivity.

Only occupational tasks involving contact with live poultry were associated with increased risks of seropositivity, and the probability of carrying H5 antibodies increased with increased numbers of such occupational contacts from around 3 percent at the low end to nearly 10 percent at the high end. A study in Italy found anti-H7 antibodies in 3.8 percent of serum samples from poultry workers during the period in 2003 when LPAI H7N3 was circulating (Puzelli *et al.*, 2005).

A study of H7N7 infection among persons reporting influenza-like symptoms was conducted in the Netherlands in association with the 2003 H7N7 HPAI epidemic (Koopmans *et al*, 2004). H7 virus was detected by PCR in ocular swabs of 86 of 453 persons (18.9 percent) calling into the health department. The highest detection rates were found in poultry cullers (41.2 percent), followed by veterinarians (26.3 percent) and farmers and their family members (14.7 percent).

Fortunately, thus far, the recent HPAI viruses do not easily infect humans, have not acquired sustained human-to-human transmissibility and only the Asian HPAI H5N1 appears to have a high case fatality in infected humans. Increased exposure of humans to avian IAVs undoubtedly increases the likelihood that avian and human influenza viruses infect the same individual, which could facilitate the emergence of a 'novel' virus, but the molecular changes required for efficient transmission of IAVs among humans remain poorly understood. Thus, in a ferret model (used to investigate the transmissibility of human influenza viruses) two H5N1 avian-human reassortant viruses and one H3N2 human-avian reassortant virus exhibited reduced replication efficiency and no transmission, suggesting that H5N1 viruses may require further adaptive steps in order to develop pandemic potential (Maines *et al.*, 2006), the likelihood of which remains unknown.

Management of Animal and Public Health Risks

Animal and public health risks have a complex relationship with economic incentives. Some responses to market and regulatory signals elevate infection risk, including the scope and term of disease incubation, while others reduce risk levels. Generally speaking, incentives for risk reduction are associated with a 'virtuous cycle' of product quality, reputation, and profit.

Behaviours that increase risk usually arise from uncertainty, loss aversion, or illicit profit incentives. Although the effects of HPAI primarily become evident in domestic poultry, disease outbreaks have repercussions that go far beyond primary producers. These repercussions are to a large extent a result of public and private responses to the (real or perceived) risk of the disease and its potential effects rather than to the actual, direct on-farm impact. Thus, any control programme needs to take into account this plurality of stakeholder reactions and interests as well as their potential to contribute to, and their incentives to undermine control programmes. It must be recognized that some individuals and enterprises may actually gain from an animal disease outbreak affecting others.

Disease risk can be considered as the outcome of an initial process of infection, followed by within farm/flock transmission, exposure of other actors in the production process, and reaction. The magnitude of each of these processes can be positively or negatively affected by economic incentives and policy interventions. For example, the risk posed by an infectious disease related to food animal production may be influenced by:

- Providing incentives and introducing regulations to promote adoption of practices that reduce the probability of initial outbreaks (e.g. bioexclusion measures);
- Introduction of incentives and standards that facilitate early detection, on-farm containment, and eradication;
- Establishing incentives and standards to reduce release from farms and other routes of exposure of others (e.g. biocontainment measures); and
- Developing strategies to mitigate disease impact, for example through emergency or preventive vaccination in high risk areas.

It would seem logical that 'private' decision makers will primarily focus on bioexclusion and less so on on-farm biocontainment, since the latter is only relevant once the pathogen has been introduced into their facilities, and as they are likely to believe that their bioexclusion measures are sufficient to prevent that from happening.

A critical issue of incentives and disincentives relates to early detection of disease incursion. The importance of early detection cannot be overstated as the magnitude of disease epidemics is exponentially related to the time elapsed between pathogen introduction and implementation of control measures. Timely reaction heavily relies on early detection and disclosure by those in daily contact with livestock, however current disease control policy tends to encourage individual behaviour along the old adage 'shoot, shovel and shut up'. Incentive systems need to be devised that encourage 'good behaviour' while penalizing 'bad behaviour'. A dilemma in the latter will be that the required disincentives may run counter to the need for incentives for early disclosure. Compensation schemes that offer equal compensation for lost animals, irrespective of timing of disclosure fail to create an incentive consistent with the policy objective of early reaction (Graming *et al.*, 2006). Likewise, depopulation of entire premises in the event of selected pathogen entry into a poultry house, for example, does not offer an incentive for reporting the latter or for major investments in within-farm biocontainment (let alone betweenfarm biocontainment).

Compensation for depopulation usually only covers (partial) costs to producers that are directly affected by the depopulation measures, while the other costs of disease control measures, such as movement restrictions, on other market participants are not normally covered. Unless an appropriate balance is struck between the control and compensation systems, farmers in a quarantined area may have an incentive to introduce disease onto farms with consequential depopulation and partial compensation rather than to suffer the indirect impacts of disease control measures.

As it can be more profitable to raise or move animals for 'finishing' to locations where animal feed is abundant, e.g. close to feed mills, than to continuously move feed over large distances, areas of high livestock density have emerged in a number of regions worldwide. Semi-vertical integration of production processes, where a large company supplies young stock and feed, while farmers provide animal housing and labour, has often not been accompanied by systematic spatial planning of the units in the system. Although spatial concentration is convenient from an organizational point of view, as illustrated in the case of the HPAI outbreaks in DPPAs, it has serious drawbacks for the control of epidemic diseases. In the European Union (EU), locationspecific disease risks, as for example determined by concentration of production units or proximity to wildlife reservoirs, are not factored into the cost of production because the current tax-financed system of disease outbreak response acts as a free insurance of last resort, and thereby results in the generation of avoidable amounts of 'risky' production (Jansson *et al.*, 2006). The same authors show that although moving to risk-based compulsory insurance for FMD, financed by the livestock sector, would not result in major relocation of dairy production within the EU, it would lead to a fairer distribution of disease control costs between member countries and between consumers and producers. Location of poultry and pig production, which does not rely on the availability of grazing land, might however shift in response to a risk-based insurance system.

The benefits of freedom from highly contagious diseases are shared by all market participants, i.e. non-exclusive, and therefore have similarities with common property resources. But as achievement and maintenance of disease freedom is heavily dependent on individual behaviour, livestock keepers may become locked into a 'Nash equilibrium' in which no one has anything to gain by unilateral action. Public intervention is required to align individual and societal interests, but to do so successfully requires detailed understanding of the individual incentives of market participants and of the full set of consequences of potential interventions.

Chapter 4

Cattle Housing

Cows play an extremely important role in most African cultures. The ownership of cattle will often be the deciding factor in a man's social position in the community because the herd may be the only practical way of accumulating wealth. However, of greater importance is the fact that cattle represent a source of high protein food, both milk and meat. This chapter focuses on housing requirements for cattle kept primarily for milk production. Little or no housing is required for herds maintained only for beef production and special handling and support facilities are discussed separately.

Much of the dairying in East and Southeast Africa occurs at elevations of 1500 metres or more. European breeds have been successfully established under these circumstances. However, European bulls crossed with Zebu cows have produced animals that are more tolerant of high temperatures than the European breeds and significantly better producers than the Zebus. Whether purebreds or crosses, they will not provide a profit to the farmer if they are left to find their own feed and water and are milked irregularly. Experience has shown that cattle respond favourably to good management, feeding and hygiene all of which is possible in a system with suitable housing.

Herd Profiles

The composition and management of cattle herds vary considerably. At one extreme, nomadic herdsmen graze their entire herd as one unit. The small holder with only a few head may keep his heifer calves for replacements or sell them. The commercial dairy producer typically has about four-fifths of his cows milking and one-fifth waiting to calve, while heifers 10 months to calving age plus calves of various ages will approximately equal the number of milkers. Mature dairy cows are bred annually and are milked for 300 to 330 days after calving. At a closer examination it will be found that several factors influence the number of animals of various categories found in the dairy herd.

In a herd of say, 24 cows, having calving evenly distributed throughout the year and a 12-month calving interval there will be, on an average, two calves born per month. The calves are normally kept in individual pens for two to three months, there is thus a requirement for four to six pens in a herd of 24 cows. However, the need for calf-pens is halved in herds where the bull calves are sold or otherwise removed from the herd at one to three weeks of age. A longer calving interval and high mortality among the calves will decrease the required number of calf-pens, while a concentration of the calving season in the herd will increase the pen requirements. If all calving is concentrated in six months of the year, the requirement of calf-pens will be doubled.

A number of cows in a dairy herd will be culled each year for reasons of low milk yield, infertility, disease, old age, etc. These cows are best replaced with young stock from their own herd, since any animals acquired from outside the farm may bring disease to the herd. Cows are commonly culled after three to five lactations, corresponding to a replacement rate of 20 to 30% per year. In herds with very intensive production there is a tendency for higher replacement rate, but it can not exceed 40%, if the heifers are obtained exclusively from the herd itself, since only about half of the calves born are female and of these some will die or be culled before first calving due to disease, infertility, etc.

The number of maturing heifers will increase with increasing age of the heifers at first calving, increased replacement percentage and a shorter calving interval. Concentrated calving may slightly increase the number of animals during some periods of the year, and will greatly affect the distribution of animals to the different age groups.

The age at first calving of heifers of European breeds is typically 24 to 27 months, while heifers of the slower maturing Zebu cattle often are 36 months or more. Maturing heifers require little or no housing facilities in the tropics. Knowledge of their exact number and distribution in various age groups during different months is therefore not as important to a building designer as to the manager of the herd.

Heifers should be introduced in the dairy herd at least a couple of months prior to their first calving to learn and become adjusted to the handling routines and feed. In loose housing systems with free stalls (cubicles) or in tiebarns this may slightly increase the need for stalls, but normally the heifer will simply take over the stall used by the culled cow, which it replaces. In herds where cows are taken to a special calving pen during calving, one such pen per 30 cows is sufficient, since the cow and her calf will spend only a few days there. However, in herds where the calving is concentrated in a short period the requirement can increase to one calving pen per 20 cows. The pen should be at least 3.3m by 3.3m.

General Housing Requirements

As has been pointed out, cattle will be more efficient in the production of milk and in reproduction if they are protected from extreme heat, i.e. temperatures of 25 to 30°C, and particularly from direct sunshine. Thus in tropical and subtropical climates shade becomes an important factor. If cattle are kept in a confined area, it should be free of mud and manure in order to reduce hoof infection to a minimum. Concrete floors or pavements are ideal where the area per cow is limited. However, where ample space is available, an earth yard, properly sloped for good drainage is adequate.

Sun Shade

With these needs in mind a shade structure allowing 2.5 to 3m: per animal will give the minimum desirable protection for cattle, whether it be for one animal belonging to a small holder or many animals in a commercial herd. A 3x7m roof will provide adequate shade for up to X cows. The roof should be a minimum of 3m high to allow air movement. If financially feasible, all the area that will be shaded some time during the day should be paved with good quality concrete. The size of this paved area depends on the orientation of the shade structure. If the longitudinal axis is east and west, pan of the floor under the roof will be in shade all day. Extending the floor approximately one third its length on the east and on the west, a paved surface will provide for the shaded area at all times.

If the longitudinal axis is north and south, the paved area must be 3 times the roof area i.e. 1/3 to the east, 1/3 to the west and l/3 underneath. Obviously this means an increase in the cost of paving. In deciding which orientation to build, the following factors need be considered:

- With the east-west orientation the feed and water troughs can be under the shade which will allow the cows to eat and drink in shade at any time of the day. The shaded area, however, should be increased to 3 to 4m^2 per cow. By locating the feed and water in the shade, feed consumption will be encouraged, but also more manure will be dropped in the shaded area which in turn will lead to dirty cows.

- With the north-south orientation, the sun will strike every part of the floor area under and on either side of the roof at some time during the day. This will help to keep the floored area dry. A shaded area of 2.5 to 3m² per cow is adequate if feed and water troughs are placed away from the shaded area.
- If it is felt that paving is too costly, the north-south orientation is the best choice in order to keep the area as dry as possible.
- In regions where temperatures average 30°C or more for up to five hours per day during some period of the year, the east-west orientation is most beneficial.

Yards

If space is severely limited and only 4 to 5m² per cow is available, then concrete paving is highly desirable. If up to 40 to 60m² per cow is available, then unpaved yards should be quite satisfactory as long as the feed and shade areas are paved and the yard is graded for good drainage. If the small holder is unable to afford an improved structure such as a shade or a paved area for feeding, then conditions can be prevented from becoming intolerable by building mounds of earth in the yard with drainage ditches between them. From 20 to 30m² per cow will keep the animals out of the worst of the mud. The soil in the mounds can be stabilized by working chopped straw or straw and manure into the surface. A number of trees in the yard will provide sufficient shade.

Deep-Bedded Sheds

In a deep-bedded system, straw, sawdust, shavings or other bedding material is periodically placed in the resting area so that a mixture of bedding and manure builds up in a thick layer. Although this increases the bulk of manure, it may be easier to handle than wet manure alone. This system is most practical when bedding is plentiful and cheap. The space requirements for various ages of animals when there is access to a yard. By designing the building to be partially enclosed on the east and west, the shading characteristics can be improved. In as much as a well drained earth floor is quite adequate, such a building will compare favourably in cost with a shaded area which is paved.

Loose Housing with Free Stalls (Cubicles)

Although simple yard and a shade or yard and bedded shed systems are entirely satisfactory in warm climates, particularly in semi-arid areas, some farmers may prefer a system with somewhat more protection. A loose housing yard and shed with free stalls will satisfy this need. Less bedding will be required and less manure will have to be removed. Free stalls must be of the right size in order to keep the animals clean and to reduce injuries to a minimum. When stalls are too small, injuries to teats will increase and the cows may also tend to lie in other areas that are less clean than the stalls.

If the stalls are too large, cows will get dirty from manure dropped in the stall and more labour will be expended in cleaning the shed area. A bar placed across the top of the free stalls will prevent the cow from moving too far forward in the stall for comfortable lying down movements, and it will encourage her to take a step backwards when standing so that manure is dropped outside the stall. The bar must, however, not interfere with her normal lying and rising movements. The floor of the stall must be of a non-slippery material, such as soil. A good foothold is essential during rising and tying down movements to avoid injury. A 100mm ledge at the back edge of the free stall will prevent any bedding from being pulled out to the alley.

The number of stalls should ordinarily correspond with the number of animals housed, except that in large herds (80 or more), only about 90% of the animals need to be accommodated at one time. Young stock may be held in yards with shade or in sheds with either free stalls or deep bedding. The alley behind the free stalls (cubicles) must be wide enough to allow the cows smooth passage and the following minimum widths apply:

Tie-Stall Sheds

Only in the case of purebred herds where considerable individual attention is given to cows can a tie-stall system be justified in tropical areas. If such a system is chosen, stalls and equipment may be purchased, in which case floor plans and elevations may be available from the equipment supplier. However, if equipment is to be manufactured locally.

Table: *Alley Widths in Conjunction with Free Stalls (Cubicles)*

Alley between a row of free stalls and a through (increase to 4.0m if there are more than 60 cows in the group)	2.7-3.5m
Alley between a row of free stalls and a wall	2.0-2.4m
Alley between two rows of free stalls	2.4-3.0m
Alley between a feed trough and a wall	2.7-3.5m

Table: *Area for Bedded Sheds and Dimensions of Free Stalls (Cubicles)*

Animal	*Age Months*	*Weight kg*	*Bedded Shed Area per Animal (m²)*		*Free Stalls Dimensions (m)*	
			A	*B*	*Length*	*Width*
Young stock	1.5-3	70-100	1.5	1.4	1.2	0.6
Young stock	3-6	100-175	2.0	1.8	1.5	0.7
Young stock	6-12	175-250	2.5	2.1	1.8	0.8
Young stock	12-18	250-350	3.0	2.3	1.9	0.9
Bred heifers and small milking cows		400-500	3.5	2.5	2.1	1. 1
Milking cows		500-600	4.0	3.0	2.2	1.2
Large milking cows		> 600	5.0	3.5	2.3	1.2

A-Enclosed and fully covered bedded shed

B-Bedded shed in conjunction with exercise yard

Table: *Tie-Stall System Dimensions (metres)*

	Cow live weight		
Stall Section	***450 kg***	***550 kg***	***650 kg***
Platform width	1.1	1.2	1.3
length[1]	1.6	1.7	1.8
Manger width	0.5	0.6	0.65
Platform slope	2-4%		

	Nose Out System	***Nose In System***
Flat manger feed alley	1.7-2.0	1.6-2.0
Feed Alley (excluding step manger)	1.2-1.4	1.2-1.4
Service alley width		1.4-2.0
Manure gutter width		0.4-0.7
depth		0.25-0.35

[1] If cows are allowed to lie with their heads over the through, otherwise add 0.4-0.5m to the length.

The tie and feed barrier construction must allow the cow free head movements while lying down as well as standing up, but should prevent her from stepping forward into the feed trough. Most types of yokes restrict the cow's movements too much. A single neck nail, set about 1 m high and 0.2m in over the merger may bruise the cow's neck when she pushes forward to reach the feed. The feed barriers that best meet the requirements are shoulder supports and the comfort stall. Note the fixing rods for the cross tie which allows vertical movements of the chain. Stall partitions should be used at least between every second cow to prevent cows from trampling each other's teats and to keep the cow standing straight so that the manure falls in the gutter.

Bull Pens

A bull pen should have a shaded resting area of 12 to 15m² and a large exercise area of 20 to 30m². The walls of the pen must be strong. Eight horizontal rails of minimum 100mm round timber or 50mm galvanised steel tubes to a total height of 1.5m and fixed to 200mm timber posts not more than 2m apart will be sufficient. The gate must be designed so that the bull cannot lift it off its hinges and there should be at least two exits where the herdsman can escape. A service stall where the cow can be tethered prior to and during service is usually provided close to the bull pen. The stall can have ramps at the sides to support the bull's front feet.

Calf Pens

Calf mortality is often high in tropical countries, but proper management and suitable housing that protects the calf from climatic stress, infections and parasites can reduce this.

Individual pens for calves from birth to 2 to 3 months of age are often built with an elevated slatted floor. This floor, which is best constructed from 37 to 50mm by 75 to 100mm sawn timber boards leaving a 25 to 30mm slat between each board, will ensure that the calf is always dry and clean. The required minimum internal dimensions for an individual calf pen are 1200 by 800mm for a pen where the calf is kept to two weeks of age, 1200 by 1000mm where the calf is kept to 6 to 8 weeks of age and 1500 by 1 200mm where the calf is kept from 6 to 14 weeks of age. Three sides of the pens should be tight to prevent contact with other calves and* to prevent draughts. Draughts through the slatted floor may be prevented by covering the floor with litter until the calf is at least one month of age. The front of the pen should be made sò that the calf can be fed milk, concentrates and water easily from buckets or a trough fixed to the outside of the pen and so that the calf can be moved out of the pen without lifting. The milk or milk substitute fed to the calf will not provide it with enough liquid and therefore it should be given fresh, clean water daily or preferably have continuous access to water in a drinking nipple. All calves, but especially those which are weaned early, should have access to good quality forage as soon as possible to stimulate rumen development. Forage can be supplied in a rack placed above the side wall of the pen. A thatched shed with six slatted floor calf pens. This construction with a feed alley will be rather expensive but can be cheaper if calves are fed from outside. Calf pens are recommended where the cows are kept in a semi-zero grazing or zero grazing system. Another system that works well is the use of individual hutches. The hutch must be thoroughly cleanedset up in a new location each time a new calf is housed in it. Plenty of litter is placed directly on the ground inside the hutch. Protection from wind, rain and sun is all the calf requires, but always moving the hutch to clean ground is the key to success.

Housing for the Small Herd

For the small holder who wants to make the very best use of his crop land and to provide his cattle with good housing that will encourage high production, a zero grazing system is recommended. The elevation and plan views of a zero grazing unit for 3 cows, 2 heifers and a young calf. Additional stalls can be added up to a total of about 10. After that consideration should be given to two milking places and a larger feed store.

Gum poles may be used instead of the cedar posts and sawn rafters, but any wood in contact with or within 5Ocm of the ground should be well treated with wood preservative. It is desirable to pave the alley, but if that is not possible, the distance between the free stalls (cubicles) and the feed trough should be doubled or tripled. A concrete pit or sloping slab in which to accumulate manure is essential. If the alley is paved, the pit can also collect urine. In fact, paving the alley not only saves space, but the value of the urine will help to pay for the paving.

The circular manure tank has a volume of $10m^3$. This will be adequate to store the manure produced during one month plus any rainfall collected in the alley. If more stalls are added the capacity of the tank will need to be increased or the interval between the emptyings shortened. A water tank to collect water from the roof can be very useful unless there is an abundant supply of water nearby.

Housing for the Medium to Large Scale Herds

For the farmer with up to about 30 cows a yard with paved shade and feed area would be suitable. The yard and feeding area may alternatively be combined with an open sided barn designed for deep bedding or equipped with freestalls and where the herd consists of high yielding cows the milking shed may be equipped with a bucket milking machine. Some farmers with up to 30 cows may even consider using an open sided tie-stall shed. In general a medium or large scale dairy unit may include the following facilities:

1. Resting area for cows: a Paved shade, or b Deep bedding in an open sided barn, or c Free-stalls in an open sided barn.
2. Exercise yard (paved or unpaved).
3. Paved feed area:

   - Fence line feed trough (shaded or unshaded), or
   - Self feeding from a silage clamp.
4. Milking Centre:
   - Milking shed or parlour, and
   - Collecting yard (part of the exercise yard), and
   - Dairy including milk store, and
   - Motor room.
5. Bull pen with a service stall.
6. Calving pen(s).
7. Calf accommodation.
8. Young stock accommodation (yard with paved shade and feed area).
9. Bulk feed store (hay and silage).
10. Concentrate feed store.
11. Veterinary facilities:
    - Diversion pen with Artificial Insemination stalls, and
    - Isolation pen.
12. Waste store:
    - Slurry storage, or
    - Separate storage of solids and effluents.
13. Office and staff facilities.

Each of the parts of the dairy unit may be planned in many different ways to suit the production management system, and the chosen method of feeding. Some require ments and work routines to consider when the layout is planned are as follows:

- Movement of cattle for feeding, milking and perhaps to pasture.
- Movement of bulk feed from store to feeding area and concentrates from store to milking shed or parlour.
- Transfer of milk from milking shed or parlour to dairy and then off the farm. Clean and dirty activities, such as milk handling and waste disposal, should be separated as far as possible.
- The diversion pen with Artificial Insemination stalls and any bull pen should be close to the milking centre as any symptoms of heat or illness are commonly discovered during milking and cows are easily separated from the rest of the herd while leaving the milking.

- Easy and periodical cleaning of accommodation, yards, milking facilities and dairy, and transfer of the waste to storage and then to the fields.
- The movements of the herdsman. Minimum travel to move cows in or out of milking area.
- Provision for future expansion of the various parts of the unit.

Milking and Milk Handling

Hand Milking vs. Machine Milking

In developed countries, where labour is scarce and expensive, machine milking has become very widespread and it is also practiced on many large commercial dairy farms in the tropics. Milking machines not only reduces labour requirement and eliminates the drudgery of hand milking, but in most cases performs a better quality milking operation than would be done by hand. However, most of the many small dairy farms in developing countries have a surplus of cheap labour and the number of cows milked at each of them is not sufficient to economically justify the installation of a machine. Furthermore, machines require power and are more expensive to purchase then the few pieces of equipment needed for hand milking. In many developing countries there is an irregular supply of spare parts and a lack of skilled mechanics.

Machine Milking gives a good quality and operates with a uniform vacuum of 275-350mm of mercury, provides a massaging effect on the teats, and is easily cleaned. The milking machine simulates nursing by the calf. Two vacuum lines lead to the teat cups. A pulsator supplies an intermittent vacuum to one line at the rate of 45 to 60 pulses per minute. The line, connected to the shell of the teat cup, causes the teat inflation (rubber liner) to alternately expand and collapse. This massaging action promotes normal blood circulation in the teat. The second line maintains a continuous vacuum on the teat and carries the milk either to a stainless steel bucket or through a pipeline directly to the milk cooler. Bucket Milking Machine is the simplest and least expensive to install, but the milk must be hand carried to the cooler.

This type of system is often chosen for the small and medium size herd and where the cows are milked on a level floor of a stable or milking shed. The labour of carrying the milk to the cooler can be avoided by installing a transfer system. This consists of a 30 litres receiving tank, including a built in filter, mounted on wheels so that it can be moved around the stable. It is connected to the cooler with a plastic hose and the milk is drawn to the cooler by vacuum from the milker pump. The hose is reeled in or out as necessary as the cart is moved around the stable. Pipeline Milking Plants transports the milk through a pipe direct from the cow's udder to the milk cooler. Pipeline milking systems are usually installed in milking parlouts where the operator stands below the level of the cows. Although they are expensive, they save backbreaking labour and are usually designed to be cleaned in place, a feature that not only saves labour but helps to ensure good sanitation. They may also be installed in stanchion or tie-stall bates but the extra pipeline needed makes the system even more expensive.

Milk Room and Cooler

Sanitation is the primary consideration in the handling of milk whether it is from one or two cows belonging to a small holder or from a commercial herd supplying milk for the city.

In either case an adequate supply of potable water is essential for cleaning the milking equipment immediately after use. Hot water (85°C) mixed with a chemical detergent is required for effective cleaning and cold water is used for rinsing. Milk should be handled in a separate area that can be easily cleaned and that is free of insects, birds, rodents and dust. The small holder producing milk only for his own household, may be able to process, curdle, or consume his milk within a short time so that cooling is not necessary. Selling milk to the public requires higher standards of sanitation and more elaborate facilities. Whether the cows are hand or machine milked, a separate milk room adjacent to the milking stalls or milking parlour is needed. This room should be well ventilated and designed with a concrete floor sloped 20mm/m to a drain and with masonry walls having a smooth, water resistant surface that can be easily and thoroughly cleaned.

Table: *Minimum Water Requirements for Parlour and Milkroom Washing*

	Hot Water, 85°C	*Warm Water, 40°C*	*Cold Water 4-10°C*
	litre	*litre*	*litre*
Hand milking equipment	10/wash		20/wash
Bucket milking equipment	20/wash		40/wash
Pipeline milking equipment	30/ wash		60/ wash
Cooling of milk in plate type milk cooler			2-3 times the amount of milk
Parlour floor wash		1/ m², day	3-6/ m², day
Milkroom floor wash		1 / m², day	1-3/ m², day
Car wash	3/car		6/car
Bulk tank wash	25-40/wash	20-30/wash	25-35/wash
Miscellaneous	20-50/day		30-100/day

Milk is strained and cooled in this room in preparation for sale. As soon as the cow has been milked the bacteria in the milk starts to multiply, but cooling of the milk to about 4° C within 2 hours will drastically reduce bacterial growth. However, proper cooling is a very difficult problem for the small scale producer. The only practical solution may be for the individual farmers in an area to bring their milk to a central collection depot for cooling immediately after milking.

On dairy farms of sufficient size and where power is available, the milk can be cooled by cold water circulated between an evaporative water cooler and a milk cooler (plate heat exchanger), through which the milk is passed until it is adequately cooled. Where milk is stored and transported in cans, cooling can be accomplished by immersing the full cans in a water-filled refrigerated cooler or by passing cold water through a coil, which is immersed in the can. The large scale dairy farm, having a pipeline milking system, and the milk collection by a road tank van, will require a refrigerated cooler and holding tank.

Milking Parlour for the Medium Scale Herd

For the farmer with 10 to 30 cows and a yard with a paved shade and feed area, the milking parlour is of suitable design. Two stands will be sufficient where the herd number is 8 to 14, but more stands should be added as indicated when the herd number increases. Hand milking would probably be used for an operation of this size. If machine milking is installed the vacuum pump and the engine, which powers it, can be put in the engine room, which is indicated in outline in the plan view. This is arranged by closing off a portion of the store room with a simple partition.

A milk cooler will be necessary to cool and hold the milk for pick up. This and facilities for washing and storing the milking equipment will be accommodated in the milk room, while concentrates are kept in the store room. A milk room should face the prevailing wind to ensure good ventilation and to keep it as cool as possible, but any openings should be screened with insect mesh.

Milking Parlour

On commercial farms where several cows are milked at the time, a milking parlour becomes a feasible investment. Several types of milking parlours are in use in dairy regions throughout the world. Any type of parlour should have a high quality concrete floor and metal railings for durability and ease of cleaning. Walls are not required, but if supplied they should at least be plastered masonry walls. The pit where the milker stands should have a floor level 900mm below that of the cattle stands for the most comfortable work position. The number of stands is determined by the allowable milking time of the herd or time taken to the concentrate ration.

Abreast Parlour

The abreast parlour allows cows to enter and leave individually. The variation of this parlour shown here, in which the front of the stands can be opened so that the cows can proceed forward out of the parlour after milking has proved effective. The main drawback with the abreast parlour is the relatively long distance to walk between milking points, and cows obstructing the herdsman, since they share the same floor space.

The stands should be 1.0 to 1.1m wide when a bucket milking machine is used or when hand milking is practiced, while 0.7 to 0.8m is adequate when a pipeline milking system is installed. In both cases the width for the milker should be 0.6 to 0.8m. A two-level abreast parlour, in which the milker works at a lower level than the cows stand, is more difficult to construct and has no outstanding advantages over the single level type. The abreast parlour has been common in East Africa for herds of more than 40 cows, but its uses is decreasing and giving way to the double herringbone parlour.

Tandem Parlour

The tandem parlour also allows for individual care of the cows. It is used mostly for smaller commercial herds and in particular, for herds with high yielding cows. The main drawbacks with this type of parlour are its larger space requirement and more expensive construction when compared to other types of parlours; of similar capacity. The parlour capacity in terms of cows milked per hour and labour efficiency can compare to that of a small herringbone parlour.

Walk-through Parlour

In walk-through or chute parlours cows enter and leave in batches. They have been used mainly for small herds. Their narrow width can be an advantage where a parlour is to be fitted in an existing building, but it is inferior to other types in most other respects, however, it is cheaper to construct than a tandem parlour.

Herringbone Parlour

The herringbone parlour layout results in a compact working area and allows feeders to be fixed to the side walls. Four stands on each side of the pit, is the minimum size of this type for high labour efficiency. If the herd has fewer than 80 cows, then a double-three parlour will keep the investment lower with only a small drop in labour efficiency. The popularity of the herringbone parlour is mainly due to its simplicity and its high capacity measured in numbers of cows milked per man-hour. (A man-hour is the equivalent of one man working for one hour). However, the risk of cows kicking the herdsman is greater in this type than in parlours where the herdsman stands alongside the cow. Double 6, 8, 10 and even 12 stand parlours are used for very large herds. These larger parlours allow more cows to be milked per hour, but because of the need for more workers and the increased waiting time to allow all cows on one side to finish before they are released, the output per man-hour is usually less.

Grain Feeders

It is advantageous to equip milking parlours with grain feeders which allow each cow to be fed in ratio to her production. Since cows are more likely to enter the parlour when they expect to be feds some labour will be saved. Manual distribution of the concentrates with a measuring scoop is recommended except in the largest herds. Semi automatic and automatic systems are expensive to install and require spare parts and mechanics for their maintenance and these may not be available when needed.

Collecting Yard

The cows are normally assembled in a collecting yard (holding area) before milking. This may be a portion of the yard that is temporarily fenced off with chains. The collecting yard should have a minimum size of 1.1 to 2.0m² per cow. Large horned cows and a low herd number will require the largest space per cow. Provision must be made for water for the cows awaiting their turn to enter the parlour. The area should slope away from the parlour 20 to 100mm/m. This not only improves drainage, but also encourages the cows to face the entrance. The collecting yard should be paved for easy cleaning and to allow for sanitary conditions in the parlour. A roof is desirable for shade and to avoid wet cows entering the parlour in the rainy season and it will reduce the amount of rainwater that has to be stored in the manure pit.

Entrance and Exit

An entrance into the parlour that is straight (no turns) will ensure a smooth and convenient operation. Once trained, cows and heifers will walk readily into the parlour. A single step of about 100mm will help to keep manure from being carried into the parlour. An exit leading into an uncrowded area will facilitate animal flow. A straight exit is desirable but not as important as a straight entry. If exiting alleys are needed they should be narrow (700 to 900mm depending on cow size), to keep the cows from turning around.

Feeding Equipment

One advantage of loose housing of cattle is the opportunity to construct the feed trough in the fence allowing easy access for filling. The simplest type of manger consists of a low barrier with a rail fixed above. However, cattle have a tendency to throw feed forwards while eating, but a wall in front, will reduce this problem. The dimensions of the trough must be chosen to conform with the height, reach and required width of the feeding space for the animals to be fed, while providing enough volume for the amount of feed distributed at each feeding time. Although timber construction is simple to install, concrete should be considered because of its greater durability. When timber is used, the base should be well treated with wood preservative. However, the preservative should not be used on any surface which cattle can reach to lick as some preservative materials are toxic to animals.

When concrete is used, it should be at least C20, or a nominal mix of 1:2:4; since a lower grade concrete would soon deteriorate due to chemical attack by feed stuffs and the cow's saliva. The cows will press against the barrier before and during the feeding so that the head rail must be firmly fixed to the vertical posts, which are immovably set in the ground. A 2.5m wide concrete apron along the feed trough will reduce the accumulation of mud.

A narrow step next to the trough will help to keep the trough free of manure as animals will not back up on to such a step. The bottom of the feed trough should be at a level 100 to 400mm above the level at which the cow is standing with her front feet. A slightly more elaborate feed trough separates the cattle by vertical rails or tomb stone barriers, to reduce competition during eating. The tombstone barrier may also reduce fodder spillage because the cow has to lift her head before withdrawing it from the trough. A simple roof constructed over the feed trough and the area where the cows stand to eat will serve as a shade and encourage daytime feeding in bright weather while serving to protect the feed from water damage in rainy periods.

Watering Equipment

Drinking water for cattle must be clean. Impurities may disturb the microbiological activities in the rumen. Table shows the requirement of drinking water, but a hot environment may considerably increase it. In dairy cows the need for water will increase with milk yield.

Cows	*Calves*	*Heifers*	*Mature*
A Reach at ground level	550	650	700
B Reach at 300mm above ground level	700	850	900
C Thought height	350	500	600
D Height to the withers	1000	1200	1300
Width of feeding space:when all			
animals feed at once	350-500	500-650	650-750
feed always available	100	150	220
Level of feed trough bottom above			
level of stand	50-200	100-300	100-400

Dimension	*Calves*	*Heifers*	*Mature Cows*
A	800-900	900-1000	1000-1200
B	300	400	500
C	50-200	100-300	100-400
D	500-700	650-850	700-900
E	300-550	400-650	450-700

Dimension	*Calves*	*Heifers*	*Mature Cows*
A	850-950	1000-1100	1100-1200
B	350	450-500	550
C	50-200	100-300	100-400
D	500-700	650-850	700-900
E	300-550	400-650	450-700
F	150-250	150-450	500
G	130-150	170-200	200

Table: *Drinking Water Requirement for Cattle*

	litre/day
Calves	10
Young stock, average	25 (8-12/100 kg body weight)
Heifers	35-45
Beef cows	30-45
Beef cattle	15-30
	(30-60 in hot environment)
Dry dairy cows	40-60
Milking cows	SO-100

Water Troughs

The size of a water trough depends on whether the herd is taken for watering periodically or is given water on a continuous basis. If water is limited, the length of the trough should be such that all of the cows can drink at one time. A trough space of 60 to 70cm should be allowed for each cow. For free choice, the trough should be sized for 2 to 3 cows at a time. One trough should be provided for each 50 animals. The length may be increased if necessary. A float valve installed on the water supply pipe will control the level automatically. A minimum flow rate of 5 to 8 litres per minute for each cow drinking at one time is desirable. To prevent contamination of the water trough with manure, the trough should preferably have a 300 to 400mm wide step along the front. The animals will readily step up to drink, but will not back up onto the raised area. An alternative is to make the sides facing the cattle sloping.

Young stock held in a loose housing system require one water trough for each 50 to 60 animals. A 60cm height is satisfactory. A minimum flow rate of 4 to 5 litres per minute for each animal drinking at one time is desirable.

Automatic Drinkers

Automatic drinkers which are activated by the animals provide a hygienic means of supplying water for cows and young stock. When used in loose housing systems for cows the bowl should be placed at a height of 100cm and be protected by a raised area beneath it. (1m^2 and 150 to 200mm heigh). One bowl should be provided for each 10 to 15 cows. A nipple drinker without bowl provides the most hygienic means of watering for young stock, but most nipples have limited flow rate and can therefore not be used for calves older than 6 months.

Feed Handling

The types and quantities of feed stuffs to be handled varies greatly from farm to farm.

Dry Hay or Forage

If an adequate supply of green forage can be grown throughout the year, then only temporary forage storage and space for chopping is required. On the other hand if a prolonged dry season makes it necessary to conserve dry forage, a storage method that will prevent spoilage is essential. A raised slatted floor with a thatched or corrugated steel roof will provide good protection for hay. A simple storage similar to the sunshade. If the store is filled gradually, it may help to have some poles in the top of the shed on which to spread hay for final drying before it is packed into the store. Loose hay weighs about 60 to 70 kg/m^3. Although requirements will vary greatly a rough guide is 3 to 5 kg of hay or other forage per animal per day of storage.

Silage

Good quality silage is an excellent feed for cattle. However, it is not practical for the small holder with only a few cows because it is difficult to make small quantities of silage without excessive spoilage. Successful silage making starts with the right crop. The entire maize plant including the grain is ideal as it has enough starch and sugar to ferment well. In contrast many grasses and legumes do not ferment well unless a preservative such as molasses is added as the forage is put into the silo. It takes a good silo to make good silage. The walls must be smooth, air-tight and for a horizontal silo the walls should slope about 1:4 so that the silage packs tighter as it settles. The forage to be made into silage should be at about 30 to 50% moisture content and must be chopped finely and then packed tightly into the silo. The freshly placed material must be covered and sealed with a plastic sheet. Failure at any step along the way spells disaster. The large commercial farmer, with well constructed horizontal or tower silos and the equipment to fill them, has the chance to make excellent feed. However, good management is no less important, regardless of size.

Concentrates and Grains

Again the amount to be stored is highly variable. The method of storing is little different from food grains and suitable storage facilities.

Manure Handling

Careful waste management is needed:

- to utilize the fertilizing qualities of the manure, urine and other wastes;
- to maintain good animal health through sanitary facilities;
- to avoid pollution of air and water and to provide good hygiene around the farmstead.

The method of disposal depends on the type of wastes being handled. Solids can be stacked and spread on fields at the optimum time of year, while liquids must be collected in a tank and may be spread from tank-wagons. Manure from a livestock production unit may contain not only faeces and urine, but also straw or other litter materials, spillage from feeding, and water. If silage is produced on the farm, the runoff from the silos should be led to the urine collection tank. Depending on the wilt the amount of effluent can vary from zero to 0.1 m^3 or more per tonne of silage but normal storage allowance is 0.05 m^3 per tonne. Manure is handled as solid when the dry matter content exceeds 25%. In this condition the manure can be stacked up to a height of 1.5 to 2 metres. This condition of the manure is only obtained when urine is drained away immediately and a prescribed amount of litter, like straw or sawdust, is used. The use of 1 to 2.5 kg of litter per cow per day will ensure that the manure can be handled as a solid. Manure with less than 20% solids has the consistency of thick slurry. It must be collected in a tank or pit but is too thick to handle effectively with pumps. It must be diluted with water to less than 15% solids before it can be pumped with a conventional centrifugal pump. If diluted in order to use irrigation equipment for spreading liquid manure, the solids must be below 4%.

The amount of manure as well as the composition varies depending upon factors such as feeding, milk yield, animal weight, position in the lactation period, and health of the animal. Cattle fed on 'wet' silages or grass produce more urine. Table shows the manure production in relation to the weight of the animals. To estimate the volume of manure and bedding, add the volume of manure from Table to half the volume of bedding used. Heavy rains requires removal of liquid for stacked manure, within the storage period. Similarly the storage capacity must be increased by about 50% or a roof should be built over the storage when slurry or liquid manure.

Example

Find suitable dimensions for a slurry manure pit with access ramp given the following:

Animals: 5 dairy cows 500 kg

Storage period: 30 days

Maximum slope of access ramp: 15%

Storage capacity (V) needed (see Table);

$V = 5 \times 30 \times 0.055 = 8.25 m^3$

Table: *Manure Production in Cattle*

Weight of animal	*Faeces*	*Urine*	*Total Manure*	*Storage capacity to be allow**
kg	*kg/day*	*kg/day*	*kg/day*	*m³/day*
Dairy cattle				
50	2.7	1.2	3.9	0.004
100	5.2	2.3	7.5	0.009
250	14	6	20	0.025
400	23	10	33	0.045
600	35	1 5	50	0.065
Beef cattle				
350	15	6	21	0.025
450	19	8	27	0.035
550	24	10	34	0.045

* These values are for manure only-no bedding is including. Washing water used in the milking parlour may amount up to 300 litres/stall/milking. Usually 50 litres/head/day to allowed.

(Normal variation can be as much as ±20% of the tabled figures).

Assume the pit will be 0.5m deep and 5m long, see sketch

Total width (W) will then be:

$W = V / (l_1 + 0\ 5l_2)\ h$

$I_2 = h / 0.15 = 0.5 / 0.15 = 3.3m$

$l_1 = l\text{-}1_2 = 5\text{-}3.3 = 1.7m$

$W = 8.25 / (1.7 \times 0.5 \times 3.3) \times 0.5 = 4.9m$

A pit 5 x 5 x 0.5m with a slope on the access ram of 15% is chosen.

Cattle Dips

Ticks continue to be one of the most harmful livestock pests in East Africa. As vectors of animal diseases ticks have been a great hindrance to livestock development especially in areas where breeds of cattle exotic to the environment have been introduced. At present the only effective method of control for most of these diseases is control of the vector ticks. Dipping or spraying with an acaricide is the mose efficient way of reducing the number of ticks.

Siting a Dip

The ground where a dip is to be built, and the area around should be slightly sloping and as hard as possible, but not so rocky that a hole for the dip cannot be dug. Laterite (murram) soil is ideal: The ground must:

- support the structure of the dip;
- be well drained and not muddy in wet weather, and
- be resistant to erosion or gullying of cattle tracks.

Cattle must not be hot or thirsty when they are dipped, so it is important to have a water trough inside the collecting yard fence.

Waste Disposal and Pollution

All dipping tanks need to be cleaned out from time to time and disposed of the accumulated sediment. It is normal for all the waste dip-wash to be thrown into a 'waste pit' that is dug close to the dip. In addition dipping tanks may crack with leakage of acaricide as a result.

The siting of the dip and the waste pit must therefore ensure that there is no risk of acaricide getting into drinking water supplies, either by overflowing or by percolating through the ground. The waste pit should be at least 50 metres from any river or stream, 100 metres from a spring or well, and considerably more than that if the subsoil is sandy or porous. A typical site layout and describes the features in the order that the cattle come to them.

Footbaths

Footbaths are provided to wash mud off the feet of the cattle to help keep the dip clean. At least two are recommended, each 4.5 metres long and 25 to 30cm deep, but in muddy areas it is desirable to have more. Up to 30 metres total length may sometimes be required. The floor of the baths should be studded with hard stones set into the concrete to provide grip, and to splay the hoofs apart to loosen any mud between them.

The footbaths should be arranged in a cascade, so that clean water added continuously at the end near the dip, overflows from each bath into the one before it, with an overflow outlet to the side near the collecting pen. Floor level outlet pipes from each bath can be opened for cleaning. If water supply is extremely limited, footbath water can be collected in settling tanks and reused later.

Jumping Place A narrow steep flight of short steps ensures:

- that animals can grip and jump centrally into the dip,
- that their heads are lower than their rumps at take-off,
- that they jump one at a time, and
- that dip-wash splashing backwards returns to the dip.

The lip of the jumping place experience extreme wear and should be reinforced with a length of 10cm diameter steel pip. The jumping place 40cm above the dip-wash level. While such a height is desirable to give maximum immersion, there could be some danger to heavily pregnant cows if the water level was allowed to fall a further 40cm. (The dipping of 1,000 cattle without replensihment would lower the water level to 60cm below the jumping place).

Splash walls and ceiling are provided to catch the splash and prevent the loss of any acaricide. The ceiling will protect a galvanized roof from corrosion. The walls can be made of wood, but masonry is most durable.

The Dipping Tank

The dipping tank is designed to a size and shape to fit a jumping cow and allow her to climb out, while economizing as far as possible on the cost of construction and the recurrent cost of acaricide for refilling. A longer tank is needed if an operator standing on the side is to have a good chance of reimmersing the heads of the animals while they are swimming, and increased volume can slightly prolong the time until the dip must be cleaned out. In areas with cattle of the 'Ankole' type with very long horns, the diptank needs to be much wider at the top.

Poured reinforced concrete is the best material to use in constructing a dipping tank in any type of soil although expensive if only a single tank is to be built, because of the cost of the form-work involved, the forms can be reused. If 5 tanks are built with one set of forms the cost per tank is less than the cost of building with other materials, such as concrete blocks or bricks. A reinforced concrete dipping tank is the only type with a good chance of surviving without cracking in unstable ground. In areas prone to earthquakes a one-piece tank is essential.

Catwalk and hand rails are provided to allow a person to walk between the splashwalls to rescue an animal in difficulty. In addition to providing shade, a roof over the dipping tank reduces evaporation of the dip-wash, prevents dilution of the dip-wash by rain, and in many cases, collects rain water for storage in a tank for subsequent use in the dip.

Draining Race

The return of surplus dip-wash to the dipping tank depends on a smooth, watertight, sloping floor in the draining race. A double race reduces the length and is slightly cheaper in materials, but a very long single race is preferable where large numbers of cattle are being dipped. Side-sloping of the standing area towards a channel or gutter increases the back-flow rate. The total standing area of the draining race is the factor that limits the number of cattle that can be dipped per hour, and the size shown in the drawings should be taken as the minimum.

A silt trap allows settling of some of the mud and dung from the dip-wash flowing back to the tank from the draining race. The inlet and outlet should be arranged so that there is no direct cross-flow. Provision must be made to divert rain water away from the dip.

Cattle Spray Race

A spray race site requires the same features as a dip site and these have already been described. The only difference is that the dip tank has been changed for a spray race. The race consists of an approximately 6m long and 1m wide tunnel with masonry side walls and a concrete floor. A spray pipe system on a length of 3 to 3.5m in the tunnel having

25 to 30 nozzles place in the walls, ceiling and floor, discharge dip liquid at high pressure and expose the cattle passing through to a dense spray. The fluid is circulated by a centrifugal pump giving a flow of 800 litres per minute at 1.4 kg/ cm^2 pressure. Power for the pump can be supplied by a 6 to 8 horsepower stationary engine, a tractor power take-off, or a 5-horse-power electric motor.

The discharged fluid collected on the floor of the tunnel and draining race is led to a sump and re-circulated. In addition to being cheaper to install than a dipping tank the spray race uses less liquid per animal and operates with a smaller quantity of wash, which can be freshly made up each day. Spraying is quicker than dipping and causes less disturbance to the animals. However, spray may not efficiently reach all parts of the body or penetrate a fur of long hair. The mechanical equipment used requires power, maintenance and spare parts and the nozzles tend to get clogged and damaged by horns.

Handspraying is an alternative method that can work well if carried out by an experienced person on an animal properly secured in a crush. The cost of the necessary eqyuipment is low, but the consumption of liquid is high as it is not re-circulated. The method is time consuming and therefore only practicable for small herds where there is no communal dip tank or spray race.

Pig Housing

Pig farming is realtively unimportant in most regions of Africa, as in most tropical countries, except China and South-East Asia. However, pig production is increasing in many tropical countries as processed pork finds an increasing market and pig production yields a relatively rapid rate of return on the capital employed. Pigs are kept primarily for meat production, but the by-products, such as pigskin, bristles and manure are also of economic importance. To some extent pigs compete with man for food, but they can also utilise by-products and wastes from human feeding.

Management Improvements

In many tropical countries pigs roam freely as scavengers or are raised in the back-yard where they depend on wastes for feed. Little attempt is made to obtain maximum In many tropical countries pigs roam freely as scavengers or are raised in the back-yard where they depend on wastes for feed. Little attempt is made to obtain maximum productivity. However, a few simple management practices can help to improve the productivity and health of these pigs. They include:

1. Fenced paddocks with shade and water where:
   - Pigs are protected from direct sun, which will cause sunburn, and sometimes sunstroke particularly with whiteskinned pigs.
   - Pigs can be fed supplementary feed secure from neighbouring pig.
   - Some basic measures to control disease and parasites are possible to reduce the often very high mortality rate and to improve the poor reproductive and growth performance and inferior quality of meat experienced in traditional pig production in the tropics. The paddock can be sub-divided into 4 to 6 smaller areas so that pigs can be moved from one enclosure to another at 2 week intervals.
   - Sows can be bred to selected sires.
2. Simple semi-covered pens constructed of rough timber with a thatch roof and floor of concrete. An earth floor can be used, but is more difficult to keep clean and sanitary. Several pens can be arranged in a row as required. The main disadvantage with this type of accommodation is the relatively high labour requirements for cleaning.
3. Wallows or sprinklers can be provided to alleviate heat stress. Being unable to sweat sufficiently pigs have a natural instinct to wallow to increase the evaporative cooling from the skin.

While such improvements have the advantage of low investment in buildings and less need for balanced feed rations, they should only be regarded as first steps in raising the general level in present primitive systems.

The raising of pigs in confinement is gradually replacing the old methods because of lower production costs, improved feed efficiency and better control of disease and parasites. Thus, the confinement system is usually advisable in circumstances where:

- good management is available;
- high-quality pigs ate introduced;
- farrowings occur at regular intervals throughout the year;
- land is scarce or not accessible all the year;
- balanced rations ate available;
- labour is expensive;
- parasite and disease control is necessary;

- the target is commercial production;
- herd size is reasonably large.

Some systems keep only part of the herd in confinement. The order of priority for confinement housing for the different classes of animals is usually as follows:

- Growing/finishing pigs (25-90 kg or more liveweight) for higher control daily gain, better feed conversions dna pasite control.
- Farrowing and lactating sows, to reduce pre-weaning mortality and for higher quality weaners.
- Gestating sows, to allow individual feeding and better control of stock.

Management Systems in Intensive Commercial Pig Production

There is no standard type or system of housing for pigs. Instead, accommodation and equipment are chooser to suit the type of management system adopted. However, there are certain similar principles and practices in most systems. These originate from the fact that most pig units will contain pigs of different ages and classes.

Farrowing-Suckling Pens

In small and medium scale intensive pig production units a combined farrowing, suckling and rearing pen is normally used. The sow is brought to this pen one week before farrowing and stays there together with her litter for 5 to 8 weeks when the piglets are weaned by removing the sow. The sow is often confined in a farrowing crate a few days before, and up to a week after birth to reduce piglet mortality caused by overlaying or trampling. Early weaning after a suckling period of 5 to 6 weeks or even less can only be recommended where management and housing is of good standard. The piglets remain in the farrowing pen after weaning and until they are 12 to 14 weeks of age or weigh 25 to 30 kg.

Group keeping of farrowing-suckling sows that have given birth within a 2 to 3 week interval is possible, but is unusual in intensive production. However, there are few acceptance problems and the litters cross-suckle and mix freely. The pen should have at least 6m^2 deep litter bedding per sow, with an additional creep area of 1m^2. In a large scale unit, which has a separate farrowing house, sometimes either of the following two alternative systems ate practiced instead of the system described above: The first alternative is similar to the system already described, but the piglets are moved two weeks after weaning to a weaner pen where they may remain either until they are 12 to 14 weeks of age (25 to 30 kg) or until 18 to 20 weeks of age (45 to 55 kg). Note that the piglets should always remain in the farrowing/ suckling pen for a further 1 to 2 weeks after the sow has been removed so that they are not subjected to any new environmental or disease stress at the same time as they are weaned. The weaning pens can contain one litter or up to 30 to 40 pigs. The pigs are often fed 'ad libitum'.

In the second alternative showing the sow is placed in a farrowing crate in a small pen one week prior to birth. Two weeks after farrowing the sow and the litter are moved to a larger suckling pen. The piglets may remain in this pen until 12 to 14 weeks of age or be transferred to weaner accommodation two weeks after weaning.

Dry Sow Pens

After weaning a sow will normally come on heat within 5 to 7 days and then at 3 week intervals until successful mating. The average weaning to conception interval can vary between 8-20 days depending on management. In the period until pregnancy has been ascertained the sow is best kept in a pen or stall in close proximity to the boar pen. Gestating sows are kept in yards or pens in groups of up to 10 to 12 sows, that will farrow within a 2 to 3 week interval. They can also be kept in individual pens confined in stalls or tethered in stalls.

Weaner and Fattening Pens

The weaners, whether they come from a farrowing pen or a weaner pen, will at 12 to 14 weeks of age be sufficiently hardened to go to a growing/finishing pen. Finishing can be accomplished either in one stage in a growing/ finishing pen from 25 kg to 90 kg-systems or in two stages so that the pigs are kept in a smaller growing pen until they weigh 50 to 60 kg and are then moved to a larger finishing pen where they remain until they reach marketable weight. In large scale production the pigs are arranged into groups of equal size and sex when moved into the growing/finishing pen. Although finishing pigs are sometimes kept in groups of 30 or more, pigs in a group of 9 to 12, or even less, show better growth performance in intensive systems. An alternative, where growing and finishing are carried out in the same facility, is to start about 12 pigs in the pen and later, during the finishing period, reduce the number to 9 by taking out the biggest or smallest pigs from each pen.

Replacement Pens

In intensive systems a sow will, on average, produce 3 to 6 litters before she is culled because of infertility, low productivity or age. Young breeding stock should be separated from the rest of the litter at about 3 months of age, since they should be less intensively fed than the fattening pigs. Gilts are first covered when they are 7 to 9 months of age or weight 105 to 120 kg. After mating they can either be kept in the same pen up to 1 week before farrowing, or kept in the gestating sow accommodation, but in a separate group. Boars in the tropics are usually quiet if run with other boars or with pregnant sows, but may develop vicious habits if shut up alone.

Determining the Number of Pens and Stalls Required in a Pig Unit

One objective in planning a pig unit is to balance the accommodation between the various ages and numbers of pigs. Ideally, each pen should be fully occupied at all times, allowing only for a cleaning and sanitation period of about 7 days between successive groups. In the following example the number of different pens required in a 14-sow herd, where 8 week weaning is practised, will be determined.

1. Determine the farrowing interval and number of farrowings per year.

Average weaning to conception interval	20 days
Gestation	114 days
Suckling period (7 x 8 weeks)	56 days
Farrowing interval	190 days

Number of farrowings per sow and year 365/190 = 1.9

2. Determine the number of farrowing pens.

The piglets remain in the farrowing pen until 12 weeks of age.

Before farrowing	7 days
Suckling period	56 days
Rearing of weaners	28 days
Cleaning and sanitation of pen	7 days
Occupation per cycle	98 days

Thus one farrowing pen can be used for: 365 / 98 = 3 7 farrowings per year.

A 14 sow herd with an average of 1.9 farrowings per sow and year requires (14 x 19)/3.7 = 7 farrowing pens.

3. Determine the number of servicing/ gestating pens.

Average weaning to conception interval	20 days
Gestation period less 7 days in farrowing pen	107 days
Cleaning and sanitation of pen	7 days
Occupancy per cycle	134 days

Thus one place in the servicing/gestation accommodation can be used for: 365/ 134 = 2.7 farrowings per year.

With a total of 27 farrowings a year 27/2.7 = 10 places would be required.

4. Determine the number of places for replacement stock.

Presume the sows on average get 5 litters, then 20 percent of all litters will be from gilts.

Rearing of breeding stock (12 to 35 weeks)	168 days
Gestation less 7 days in farrowing pen	107 days
Cleaning and sanitation of pen	7 days
Occupancy per cycle	282 days

About 30% more animals are separated than the required number of gilts thus the required number of places in the 14 sow herd will be (14 x l.9 x 0.2 x 1.3 x 282) / 365 = 6 places

5. Determine the number of places in the growing/finishing accommodation:

One stage finishing:

Fattening of pigs 12 to 27 weeks of age, (25-90 kg)	105 days
Extra period for last pig in the pen to reach m arketable weight	21 days
Cleaning and sanitation of pen	7 days
Occupancy per cycle	133 days

Assuming that 8 pigs per litter will survive to 12 weeks of age the number of places required in the finishing accommodation will be: (14 x 1.9 x 8 x 133)/ 365 = 78

That is 8 pens with 10 pigs in each or 10 pens if each litter should be kept together.

Two stage growing/finishing unit:

Growing pigs 12 to 20 weeks of age will occupy a growing pen for 63 days including 7 days for cleaning.

(14 x 1.9 x 8 x 63) / 365 = 37 places is required in the unit.

Finishing pigs 20 to 27 weeks of age will occupy a finishing pen for 70 days including 14 days emptying period and 7 days for cleaning. (The empyting period will be shorter if the pigs are sorted for size while being transferred from the growing to the finishing pens. (14 x 19 x 8 x 70) / 365 = 41 places is required in the unit

From the above example it will be appreciated that the number of pens of various kinds required in a pig unit is based on a number of factors. It is, therefore, not possible to lay down hard and fast rules about the relative number of pens and stalls. However, a guide line to the requirement of pens in herds with average or good management and performance in tropical conditions.

Space Requirement

In intensive pig production systems all pigs should be raised on concrete floors to provide for a clean and sanitary environment. In semi-intensive systems a concrete floor is only used in the pens for finishing pigs and perhaps in the farrowing pens, whereas an earth floor or deep litter bedding is used in other pens and yards. Litter may or may not be used on a concrete floor, but its use is desirable, particularly in farrowing pens.

Because of the cost of a concrete floor there is a tendency to reduce the floor area allowed per animal. However, too high stocking densities will contribute to retarding performance, increasing mortality, health and fertility problems and a high frequency of abnormal behaviour thus endangering the welfare of the animals. Increasing the stocking density must be accompanied by an increased standard of management and efficiency of ventilation and cooling. In particular, to aid in cooling, finishing pigs kept in a warm tropical climate should be allowed more space in their resting area than is normally recommended for pigs in temperate climates. The recommended space allowance per animal at various stocking densities. The figures listed for high stocking density should only be used in design of pig units in cool areas and where the management level is expected to be above average.

The dimensions of a pen for fattening pigs are largely given by the minimum trough length required per pig at the end of their stay in the pen. However, the width of a pen with low stocking density can be larger than the required trough length. This will reduce the depth to 2.0 to 2.4m, and thus the risk of having the pigs create a manure are within the pen. Furthermore, the flexibility in the use of the pen will increase and the extra trough space allow additional animals to be accommodated temporarily or when the level of management improves.

Sometimes finishing pens are deliberately overstocked. The motive for this is that all pigs in the pen will not reach marketable weight at the same time and the space left by those pigs sent for slaughter can be utilized by the remainder. Such over-stocking should only be practiced in very well managed finishing units.

General Requirements for Pig Housing

A good location for a pig unit meets the following requirements: easy access to a good all-weather road; welldrained ground; and sufficient distance from residential areas to avoid creating a nuisance from odour and flies. An east-west orientation is usually preferable to minimize exposure to the sun. Breezes across the building in summer weather are highly desirable.

A prevailing wind during hot weather can sometimes justify a slight deviation from the east-west orientation. Ground cover, such as bushes and grass, can reduce reflected heat considerably, and the building should be located where it can most benefit from surrounding vegetation. A fairly light well drained soil is preferable, and usually the highest part of the site should be selected for construction.

Pig houses should be simple, open sided structures as maximum ventilation is needed. A building for open confine merit is therefore essentially a roof carried on poles. The roof supporting poles are placed in the corners of the pens where they will cause least inconvenience. A free span trussed roof design would be an advantage but is more expensive.

In some circumstances it may be preferable to have solid gable ends and one tight side to give protection from wind or low temperatures, at least for part of the year. If such walls are needed they can often be temporary and be removed during hot weather to allow maximum ventilation. Permanent walls must be provided with large openings to ensure sufficient air circulation in hot weather. If there is not sufficient wind to create a draught in hot weather, ceiling fans can considerably improve the environment.

The main purpose of the building is to provide shade, and therefore the radiant heat from the sun should be reduced as much as possible. In climates where a clear sky predominates, a high building of 3m, or more, under the eaves gives more efficient shade than a low building. A wide roof overhang is necessary to ensure shade and to protect the animals from rain.

A shaded ventilation opening along the ridge will provide an escape for the hot air accumulating under the roof. If made from a hard material the roof can be painted white to reduce the intensity of solar radiation. Some materials such as aluminium reflect heat well as long as they are not too oxidized. A layer of thatch (5cm) attached by wire netting beneath a galvanised steel roof will improve the microclimate in the pens.

***Table:** Dimensions and Area of Various Types of Pig Pens*

	Units	***Stocking density***		
		Low	***Medium***	***High***
A. Farrowing/ suckling pen.				
Resting area, if weaner pens are not used	m^2	10.0	7.5	6.0
Resting area, if weaner pens are used	m^2	8.0	6.0	5.0
Manure alley width	m	1.7	1.5	1.3
Farrowing pen (System IV)	m:	-	4.5	4.0
Farrowing crate, length excl. trough	m	2.0	2.0	2.0
width depending on size of sow	m	0.65-0.75	0.6-0.7	0.55-0.65
free space behind the crate	m	0.4	0.35	0.3
Piglet creep (incl. in resting area)	m^2	2.0	1.5	1.0
B. Boar pen				
1. Pen with yard				
Resting are (shaded)	m^2	6	5	4.5
Yard area (paved)	m^2	12	10	08
2. Pen without yard	m^2	9	8	7
C. Gestating sow pens				
1. Loose in groups of 5-10 sows				
Resting area (shaded)	m^2	2.0	1.5	1.1
Yard area (paved)	m^2	3.5	3.0	2.5
Feeding stalls, depth x width	m	2.0 x 0.6	1.8 x 0.55	1.7 x 0.5
2. Individual stalls with access to manure alley, length of stalls excl. trough	m	2.2	2.1	2.0
width of stalls	m	0.65-0.75	0.60-0.70	0.55-0.65
width of manure alley	m	1.5	1.4	1.3
3. Confined in individual stalls length x width of stalls	m	2.2 x 0.70	2.1 x 0.65	2.0 x 0.60
D. Weaner pen (to 25 kg or 12 wks)				
Resting area excluding trough	m^2/pig	0.35	0.30	0.25
Manure alley width	m	1.0	1.0	1.0
E. Growing pen (to 40 kg or 17 wks)				
Resting are excluding trough	m^2/pig	0.5	0.45	0.40
Manure alley width	m	1.1	1.1	1.1
F. Finishing pen, resting area excl. trough				
For porkers to 60 kg or 21 wks	m^2/pig	0.70	0.60	0.50
For beaconers to 90 kg or 27 wks	m^2/pig	0.90	0.75	0.60
For heavy hog to 120 kg or 33 wks	m^2/pig	1.0	0.85	0.70
Manure alley width	m	1.2-1.4	1.2-1.3	1.2

The pen partitions and the 1 metre wall surrounding the building, which serves to reduce heat reflected from the surrounding ground, can be made of concrete blocks or burnt clay bricks for durability or perhaps soil-cement blocks, plastered for ease of cleaning. Regular white washing may improve the sanitary conditions in the pens.

Doors have to be tight fitting and any other openings in the lower part of the wall surrounding the building should be avoided to exclude rats. Apart from stealing feed and spreading disease, large rats can kill piglets.

For all types of confinement housing a properly constructed easily cleaned concrete floor is required. Eighty to 100 mm of concrete on a consolidated gravel base is sufficient to provide a good floor. A stiff mix of 1:2:4 or 1:3:5 concrete finished with a wood float will give a durable non-slip floor. The pen floors should slope 2 to 3% toward the manure alley and the floor in the manure alley 3 to 5% towards the drains.

Chapter 5

Farm Business Management

Introduction

This chapter examines the management structure of farms to ascertain who controls the use of farm assets, including land and water. Management units that make decisions for farms are described, extending information about how farmers control and guide their businesses. The chapter also examines decisions of farmers from the perspective of how production, marketing, finance, and human resources are used to form farm businesses.

Characteristics of Farm Businesses' Managers

A farm's management unit consists of the individual or group responsible for decisions about how a farm will be operated. How a farm is legally organized is often viewed as being the same as its management. A proprietor makes decisions for proprietorships, partners for partnerships, and elected directors and officers for corporate farms. However, a farm's management unit may not be synonymous with its ownership. For example, land owners may or may not participate in management decisions. The Census of Agriculture reported in 1999 that 14 percent of landlords either made or shared in decisions related to selection of fertilizer and chemicals, while 13 percent helped decide cultivation practices.

Legal organization, while helpful in indicating a farm's governance structure, may not reveal who participates in farm management. Even on proprietor farms, more than one person may participate in management decisions. Of the 2.1 million U.S. farms reported by the Census of Agriculture, over 121,000 reported three or more operators.

Together, farms reported 2.7 million operators. In 2003, the primary operator made crop decisions on 52 percent of farms, two operators made joint decisions on 12 percent of farms, and a third operator was involved on 1.3 percent of farms. For the remaining 33 percent of farms, crop production was likely not a part of production activities. In the last decade, involvement by persons other than the primary operator has remained an important aspect of farm management.

Operator and operator-spouse management teams controlled 59 percent of farms in 2003. When paid or informal advisors are considered, the share rises to 89 percent. These farms were, by far, the smallest in terms of acreage and value of production. Management that featured more than two people or outside hired/informal assistance operated larger businesses. Nearly a third of these farms were commercial farms, versus about 3 percent for operator-only units and 7 percent overall. Each management structure that included outside advisors represented a disproportionate share of commercial-size farms.

Operator-only management units had the highest average age, nearly 2 years older than the average for all primary operators. Primary operators in multiple-operator management units were youngest, age 54 on average. Primary operators that used outside assistance had the largest share of college-level attainments. This group was followed by multiple-operator teams and operators in operator-only units.

Farms managed by operators only or by a combination of operators and spouses were more common in the Eastern Uplands, Southern Seaboard, and Mississippi Portal. These regions have a larger share of smaller farms run by operators who work off-farm. Farms run by multiple-operator teams and that used outside assistance were more common in the Heartland, Northern Crescent, Northern Plains, Fruitful Rim, and Basin and Range. Multiple operators were more common on farms that specialized in cash grains and soybeans, high-value crops, and dairy.

Farm Business Management Entails a Host of Choices

Managers of farm businesses make choices about inputs and their use in producing crops, livestock, or other products and services. Production decisions focus on whether to produce crops, livestock, both, or nothing (for example, by placing land in conservation). Financial decisions centre on acquiring and maximizing the use of inputs. Do managers have sufficient funds to buy inputs like seed or fertilizer (short-term decisions) or invest in capital items like equipment? If not, is borrowing warranted? Marketing options range from cash markets to contracts to direct sales (farmers' markets, the Internet, wholesale/retail buyers, or livestock producers.)

Human resource issues include the amount and timing of labour needed to undertake production. In 2003, 45 percent of operators reported their primary occupation as other than farming. An even larger share worked off-farm. Thus, work arrangements vary from self-sufficiency to inclusion of household members, other operators, and a variety of custom hire, contract, and hired workers. Farmers may even work off-farm and hire someone else to do farm work.

Classifying Farm Business Systems

The result of all the choices across all these business concerns is a highly diverse farm sector. Some farms amount to a single individual supplying all labour to produce one, or maybe even no commodities, with cash sales, and without debt. Other farms produce multiple commodities, market to various outlets, use a variety of labour sources, and take on debt from multiple lenders structured for different periods of maturation.

One way to overcome this complexity empirically is to devise a classification system that jointly considers management choices. To develop the farm business system classification, each of four business areas—production, finance, human resources, and marketing/contract use—was measured as a dichotomous variable. For example, farms producing 2 or fewer commodities were assigned a score of zero, while those with 3 or more commodities were given a score of 1. The same scoring convention was used for debt, hired labour, and cash sales versus production or marketing contracts. By equally weighting each of the four business activity areas, a total score ranging from 0 to 1 was calculated to reflect the overall complexity of the operation. For example, a score of 0 indicates a farm having two or fewer commodities, no debt, operator/family labour only, and cash sales.

Characteristics of Farm Business Systems

The 31 percent of farms with the least complex farm business system controlled 11 percent of acres operated and generated less than 3 percent of production value in 2003. Over 85 percent of these farms are rural residences and the rest almost entirely intermediate farms (sales below $250,000 and the operator reports farming as his or her major occupation). Over 98 percent had sales of less than $100,000. These farms specialized in production of field crops other than cash grains or soybeans, beef cattle and general livestock. Operator and operator-spouses managed three-fourths of the least complex farms.

They had the highest average operator age and the largest share of primary operators over age 65 years (32 percent). Over 80 percent considered their primary occupation to be off-farm and almost 29 percent were retired. This helps explain the 890 hours worked onfarm by the operator, well below the all-farm average of 1,393 hours.

The most complex farms controlled 8 percent of acres and generated 18 percent of value of production in 2003. Farms in this group were mostly commercial. Over 27 percent had over $500,000 or more in sales (versus 3 percent of all farms). The most complex farms had a much larger share of management teams that included multiple persons. These farms were more common in the Northern Crescent, Heartland, Northern Plains, and Eastern Uplands. They tend to specialize in dairy, poultry, hogs, cash grains, and soybeans. Nearly two-thirds reported hiring other individuals and three-fourths had custom hire assistance. The primary operators in these management units were younger, averaging 49 years, nearly 7 years less than the all-farm average. On average, primary operators reported working over 3,000 hours on their farms in 2003, with spouses and other operator labour adding more than 1,100 hours to the total.

Summary

Farm managers not only have to be highly skilled at the technical aspects of farm production, but they also have to handle primary and support activities for their farms that range from input procurement to technology, finance, accounting, and human resource management (Gray et al.). Changes in crop and livestock production, including use of contract arrangements and technologically modified seed stock, mean that managers may need to interact more with both suppliers and customers. With the rising cost of inputs, particularly capital items such as machinery and equipment, managers also have to control a range of financial arrangements that transcend farm mortgages. Even land rents have become more complex, with some arrangements incorporating changes in prices and yields.

As a farm grows, expertise to handle these tasks either has to be available within the existing owner-operator-management arrangement or be acquired by adding to the management team. Survey results suggest that the size and composition of management teams align with the complexity of farm businesses. The least complex farms were most often managed by a single operator or by a combination of operator and spouse. Conversely, the most complex businesses typically involved operators, spouses, other partners, and outside advisors. Many Federal and State programs provide income and technical/other assistance to farmers, specifically a farm's decisionmaker. In today's farm sector, that person is not automatically the farm operator alone, especially on farms with the most cropland and production.

Farm Management

Farm management deals with the organisation & operation of a farm with the objective of maximizing profits from the farm business on a continuing basis. The farmer needs to adjust his farm organisation from year to year to keep abreast of changes in methods, price variability & resources available to him. Thus farm management is the science which deals with the analysis of the farming resources, alternatives, choices & opportunities within the framework of resource restrictions & social & personal constraints of farming business. This complex information is integrated and synthesized to increase profitability of the farming business, the ultimate aim being to raise the standard of living of the farming people. This does not mean that farm management deals exclusively with the maximization of income; in fact, it takes into account the goals and objectives of the individual farmer, other than income maximization. Thus this discipline deals with people or organisers and decision-makers in respect of farms and agricultural production. It is people-oriented rather than crops or livestock per se.

Farm management is a decision-making science. It helps to decide about the basic course of action of the farming business. The basic decisions of the farming business are:

(a) What to produce or what combination of different enterprises to follow?

(b) How much to produce and what is the most profitable level of production?

(c) What should be the size of an individual enterprise, which, in turn, will determine the best overall size of the farm business?

(d) What methods of production (production practices or what type of quality of inputs and their combination) should be used?

(e) What and where to market?

Management of a Farm

Farm management is different from what is commonly confused with the work of a farm manager who manages a government farm as an agronomist. His function is normally limited to supervising & handling the day-to-day routine of a farm. It normally pertains to the existing pattern of the resource-use & crops-mix under which only the existing plan is executed, supervised & carried out. An intelligent farm manager may go a little further & look after the farm machinery to keep it going.

Farm management, is however, much more than that. Here we are not just concerned with the distribution of labour & irrigation water for day-to-day operations. The emphasis is on the decision-making function of evaluating & choosing between alternative strategies. A major concern is about adjustments which are more suitable & profitable & about exploring new situations & opportunities for maximization of income & satisfying other goals of a farmer. It is the approach under which the opportunity costs of the various resources are evaluated & adjustments in resource-use & enterprise mix are made to secure higher levels of farm income.

Following this approach, the emphasis on yield (productivity per unit of a resource, mainly land) is not ignored, but the greater focus is on the increasing of farm income through a second business organisation. As a business, farming requires the application of business methods & efficient management. To be an efficient manager, one must keep oneself abreast of developments in new technology, new practises, price trends, & economic outlook. Again a farm management man should identify the constraints in the external environment conditions, which hamper a farmer's opportunities & plans for making adjustments in his farm organisation & render the technically superior production plan economically unattractive to the farmer.

Farm Management and other Sciences

Basically farm management is economics in the context of fundamental definition of economics which involves three elements, viz. the scarcity of resources. Their alternative uses & the objective of profit maximization. It is this science which deals with making rational, profitable & economic recommendations to the farmers. It also deals with the growth & stability aspects of economics.

The basic information about the multiple (alternative) uses to which the scarce resources can be allocated is supplied by other physical & biological sciences. The research in these sciences should continually generate the relevant data on alternative technologies & practices whose profitability can be tested under actual farm situations. And these data are required for the farm organisation as a whole & not far for a single hectare, animal or tree.

The challenge to the farm management specialist is to be able to integrate & synthesize the diverse pieces of information from many disciplines such as agronomy, animal husbandry, horticulture, soil science, plant breeding, entomology, plant pathology, general economics, sociology & psychology, into an optimum 'package' which can be used on the farms with profit. Agricultural scientists mainly put emphasis on the maximisation of yield rather than on the use of the optimum level of resources. But the goal in farming is not to make a profit on some single enterprise or from a part of the farm land as some *krishi pandits* do, but to use land, labour, & capital resources in such a way that they make the greatest contribution to the total profits from the entire farm.

The superiority of this discipline, thus, lies in its treating the farm as an operational unit & tailoring the recommendations of all other disciplines to fit into an individual farmer's pattern of resources.

In the context of the recent technological breakthrough, management today should be viewed as a process within a rapidly moving frame of reference. " It is now more scientific, less artistic; more dynamic, less static; more versatile & less rigid". Farm management is forward-looking in its approach. Its task is not so much the improvement of the present farming practices but of the establishment of the whole sets of new production methods & farming systems which would put our agriculture on a continuously rising growth curve.

Government policies should change the economic environment to help the interests of the farmers to converge on the national goals. Studies on farm management to determine the responsiveness of the farmers to different levels of prices become extremely relevant in inducing the farmers to produce the quantities of different agricultural products & services needed by society. A rational pricing policy for water is needed. Canal water, for e.g. is charged for without any direct relation to the quantity supplied. As a result, there is no incentive to the farmer to allocate water in the most economical manner. Farm management helps to identify such uneconomic practices & the most limiting factors. Irrigation or power (the bullock versus the tractor) may be a more important restriction than the limits imposed by land. In areas where water rather than land is the principal limiting factor & the marginal-value productivity of irrigation water is very high, it should be most carefully used rather than over irrigating a few hectares of crops needing high water consumption.

Good farm management can lead to a highly productive use of farm resources & can avail itself of the technological revolution now going on in our agriculture.

Application of Farm Management

As against the general contention that farm management is mostly applicable to crop farms, its application to other types of farming business, such as orchards, dairying, poultry, etc. is no less important. Faculty farm management has contributed more to citrus decline than the poor physico-chemical conditions of the soil, the stock-scion incompatibility, insect pests, disease & viruses. Intercropping with exhaustive crops seems to be one of the major reasons for the poor health of the plants. Citrus die-back or decline does not seem to be a disease by itself, but a mere symptomatic expression of physiological imbalance owing to the cumulative effect of several factors. Similarly, if milk production has to be encouraged, it has to be made more profitable than cereal production. One has to demonstrate how the addition of mulch animals to a farm fits into its resource organisation & yields higher returns.

With land as a scarce resource, efforts are made to intensify its use for maximising profits through multiple cropping. During a given period, the relative profitability of different crop rotations is evaluated & finally selected for adoption. 'Kufri Sinduri' variety of potato is not being adopted on a large scale in spite of its giving 10 percent higher yield than some other varieties, e.g. 'Up-to-date'. The reason is simple. The former is a late variety & delays the sowing of wheat which decreases its yield. It is not 'Kufri Sinduri' potatoes versus 'Up-to-date' potatoes; rather, it is 'Kufri-Sinduri' followed by wheat versus 'Up-to-date' followed by wheat that counts in farm management. Similarly, in West Bengal, where jute has been competing with paddy, the situation is changing with the introduction of high-yielding varieties of wheat, which can be rotated with paddy. The comparison in terms of the relative profitability now runs between jute & paddy plus wheat, & not just between jute & paddy. Owing to the paddy-wheat rotation becoming more profitable, jute production in West Bengal is suffering a set back. As Indian agriculture becomes more productive with the passage of time, the farmers will be confronted with the task of making the best selection from among many alternatives.

Scope of Farm Management on Small Farms

Sometimes it is said that management is not important on small farms. This notion is widely prevalent under those situations where the farms operate on a low level of technology. In the case of small farms, farm organisation cannot be changed as much as on a large farm, but the choice in respect of farm practices & methods of production, cropping intensity, etc. offer worthwile alternatives. Japan's method of improving agriculture based on the principles of farm management clearly demonstrates that a small farm by itself should not be a hindrance to increased production & higher income. Small farms may have different problems, but the principles of the efficient use of available resources to obtain the maximum economic returns & family satisfaction remain the same in both the cases. The principles of farm management, since they deal with the allocation of resources, apply to small & large farms equally. A Chinese proverb is very appropriate to quote:" Although the size of a sparrow is small, yet it is physiologically as perfect as a big bird".

Farm Planning

Planning means taking decisions in advance. It stimulates thinking, broadens understanding & challenges the farmer to move forward. It is a forward-looking approach. The farm plan helps a farmer to decide how to combine new ideas & old ones to his best advantage. By identifying his credit & supply needs, the farm plan helps him to arrange for the timely supplies of credit, seeds, fertilisers, etc. A specific farm plan setting fort his expected output, expenses & income, serves as a sound basis on which a credit institution can give him production credit, based on his productive capability rather than on his net financial assets. It is out of his income & not through the sale of assets that the cultivator has to pay off his loan. Thus the farm plan or the budget is to the farmer what the blue-print of the architect is to a building contractor. It shows what is to be done & how it is to be done. It furnishes an organised & logical approach to his problems & helps him to work out the solution.

Partial Budgeting Distinguished From Full Budgeting

1. Partial budgeting considers a few alternatives which do not affect the organisation vitally, but full budgeting takes care of all the alternatives.
2. In full budgeting, the inventory of the farm, the resource structure, the existing resource use & such problems as overstocking or understocking of resources, etc. are considered, but in partial budgeting information with respect to a few alternatives is considered.
3. Partial budgeting does not indicate the break-even point as to when to start one practice & abandon another, but full budgeting does.

The best strategy is to make an effective use of the partial budget at different stages of full budgeting.

Limitations of Partial Budgeting

The partial budgeting analysis suffers from some limitations, viz. it does not consider all the alternatives open to a farmer within the restraints of his present resources. It only considers a few alternatives & these too, should not affect the farm organisation to a great extent.

Partial budgeting does not always provide a complete solution. As one may demonstrate that mechanical threshing costs less per quintal than the traditional *phalla* method, but this would not serve the purpose. The other questions such as the break-even point of the thresher, i.e. the minimum volume of business (i.e. 1,000 quintals of wheat to thresh) the possibilities of custom work, the capital available to purchase the thresher, etc. should also be considered. Similarly, the question whether to increase the number of dairy animals may be answered by using partial budgeting. But, in this case the partial budgeting analysis will approach very closely to a change in the complete farm organisation. Again, the partial budget may show American cotton to be less profitable than hybrid maize, but when full budget is prepared, both the activities may not enter the farm organisation.

Like any other farm-management analysis, the result of partial budgets are subject to variations in the input-output prices & field conditions such as soil type, soil fertility, etc. Also the restraints pertinent to a particular farm organisation necessitate the tailoring of these results to ensure their applicability to means for making profits.

Complete planning & budgeting. In preparing complete plans for the farm, all the sophisticated analysis of studying an individual cultivator's opportunities, constraints & problems is done. Every little aspect of the farm organisation is examined & then suitable adjustments are studied & a suitable combination as fits in rationally with that given farm organisation is suggested. Various recommendations in terms of augmenting resources, if the need may arise, are also made. It implies the following steps.

1. Farm Map. The farm is carefully mapped out, giving its salient features, like soil type soil-fertility & rotations followed. Low-lying areas or other such features are also shown in the map. Then based on the previous crop history, land-capability classification is done & is also shown in the map.

2. Inventory of Farm Resources. Every asset on the farm, ranging from hand-tools to sources of power, etc. are inventoried. It does not provide us with the picture of the resources as owned by a farmer, but we can also work out their use-patterns & their condition, i.e. whether they will have to be replaced or whether they will be sufficient for the new plan or some augmentation will be needed.
3. Examining the Existing Organisation having prepared an inventory of the existing resources & their availability, what we are interested in is their use-pattern within the framework of the existing crop mix, whether the resources are understocked or overstocked. A careful analysis of the restrictions & weaknesses of the farm organisation is made. The weaknesses maybe many, such as:
   i. Lumpy units may not be fully utilised.
   ii. Irrigation may not be adequate.
   iii. The peak-season labour requirements may not be met.
   iv. Short-term capital may be inadequate.

It is not simply the weaknesses that are emphasised but the good points of the existing organisation are also to be highlighted. The existing organisation may be improved or new good items maybe adopted.

Laying Down Restrictions & Planning. (a) Restrictions maybe with regard to bullock-power availability, or in respect of putting some area under cotton, or vegetables for home-consumption, though such a change in the cropping pattern may not be profitable. (b) Labour-requirements, power requirements, capital needs & new equipment needed are worked out & a suitable crop mix is adopted.

In complete farm planning, the goals of the farmer are also important. He maybe desirous of investing in farming &, at the same time, he maybe willing to construct a new house. His goals & opinions & risks are all studied & weighed, & a compromising decision maybe taken. Thus in a complete farm-planning, due consideration is played to resource use, restrictions, importance of relationships among different enterprises & to the goals & managerial skills of the operator.

Farm Credit

Analysis for credit. A banker's plan is something more than the farm management specialist's plan. A banker is more concerned with the recovery of the loan advanced by him to the farmer. A farm-management specialist's plan is all right, but a banker, if he is to call it his plan, must examine it, as not only the additional returns from the alternative plan, but also whether the farmer would be able to repay the loan from the additional returns or not. The credit forms the total capital requirement minus the farmer's owned funds. Most of the banks expect the farmer to contribute 25% of the total capital requirement as margin money so that he has some stake in the business.

Season-wise analysis. Because the loan is to be paid at the end of every season, repayments include:

(a) The total amount in the case of self-liquidating loans which are for short periods & for purchasing seeds, fertilisers, hiring casual labour, etc.

(b) The instalments of the medium & long-term loans which are not self-liquidating & are for constructing buildings, raising irrigation structures, purchasing implements, machinery etc.

Farm-efficiency Measures

Efficiency is the ratio of the output to the input. This concept is important as it shows how much profitable the farming business is. Various measures to explain the efficiency of the business as a whole, & in parts, are available. When examined together they help to point out the weaknesses in the farm business & provide a guideline as to which part of the business deserves special attention for making improvements. In a particular situation due importance is given to a particular measure.

For instance different measures would be adopted for indicating the volume or size of the business, aggregate earnings from particular factors or the business as a whole & returns per unit of a particular factor input. Further, the efficiency of a farm can be judged from the costs or returns or both. No single efficiency measure is so complete as to give a true picture of the entire farm business.

These efficiency measures help to reorganise the same farm for which they are calculated but they must be used with great caution, while comparing different farms. This is especially important in the developing countries where farming is of diversified nature & individual farm business has wide variations in respect of soil type, resource restraints, capacities & capabilities of the farmers to undertake risks, their attitude towards innovations, etc.

Also, these measures suffer from limitations for making comparisons because of their changing prices, costs, & conditions of the farm business. Adjustments with the price & coat indices are necessary before such comparisons give any valuable information.

Measuring the Size of Business

Farm income is directly related to the size of the business but in no way it indicated whether the farming business is operating efficiently or not. The following are some important measures of the size of a business.

(1) Total area of farm & the number of livestock. The acreage as a measure of the size of the farm may be either the total land, or the area under crops. The acreage at its face-value carries no meaning. It is necessary, therefore that land be homogenised & standardised for variations in fertility status, irrigation & lack of irrigation, etc. by such standard measures as rental value, land revenue, etc. As in a given type of farming area, there are important crops common to all farmers; therefore, the area (standardised) under principal crop enterprises will help to introduce further refinements in this measure.

In the case of livestock farms, the number of animals can be used as a measure of the size of the business. again, the breed differences between the herds of two farms make a great difference in the returns received by them.

On farms having both crop & livestock enterprises, any of the two will not give the relevant size of the business. Both may be used to represent the situation, but it is not comparable.

(2) Total capital managed. The total capital managed is obtained by adding all the capital values of inventories, including land, either of the closing & beginning inventories can be used. This measure is useful when all the farms are producing the same product & this has no meaning for different specialisations. One is also not sure at the face value of the capital investment whether it is nearer to the optimum required, one farm can be compared with the other for the size of business by the capital investment only if the two farms have the near-optimum investments. The difficulty here is that many farms are understocked or overstocked & the comparison between them is meaningless.

(3) Total inputs. The level of inputs in a broad way determines the level of output (income). The total inputs mean the operational expenses plus the fixed expenses-the rental value of land, the value of labour (of family & operator) & the interest on the

capital. This is the most reasonable method. It can be used in similar situations to compare with the assumption that rates of rent, interest & wages would be the same for each comparable situation.

(4) Gross income. This relates to the farm output during the year. This is obtained by adding the home-consumed farm products (in Rs) & the increase or decrease in inventory to the gross receipts of the year. This is a good measure but its use is limited to the same type of farming.

(5) Productive manpower units. They refer to the total manpower input required for normal efficiency. This method is more appropriate where labour is scarce & is being used optimally. In a labour-surplus situation, this method has only limited value, also, the measure is limited to the comparing of the similar type of farming situations. Again, the difference arises from the difficulty to see whether the farmers under comparison use the scarce labour where it adds to the maximum or not.

These measures are rough indicators of the size of the business & can be used to compare only similar situations at a point of time & for the same type of farming.

Cost Concepts

There are many cost concepts which are in frequent use in the farm management literature, i.e. Cost A_1, Cost A_2, Cost B & Cost C; working or variable or operational cost & fixed costs; machine costs, labour costs, livestock costs, crop costs, etc. a brief discussion of these cost concepts is given below:

(i) Cost A_1: It includes the following 16 items of costs:

1. Value of hired human labour (permanent & casual)
2. Value of owned bullock labour
3. Value of hired bullock labour
4. Value of owned machinery
5. Hired machinery charges
6. Value of fertilisers
7. Value of manure (produced on the farm, & purchased)
8. Value of seed (both farm produced & purchased)
9. Value of insecticides & fungicides
10. Irrigation charges (owned & hired tubewells, pumping sets etc.)
11. Canal-water charges

12. Land revenue, cesses & other taxes
13. Depreciation on farm implements (both bullock-drawn & worked with human labour)
14. Depreciation on farm buildings, farm machinery & irrigation structures.
15. Interest on the working capital
16. Miscellaneous expenses(wages of artisans, cost of ropes & repairs to small farm implements)

(ii) Cost A: It is cost A_1 plus

17. Rent paid for leased land

(iii) Cost B: It is cost A_2 plus

18. Imputed rental value of owned land (less land revenue paid thereupon)
19. Imputed interest on owned fixed capital (excluding land)

(iv) Cost C: It is cost B plus

20. Imputed value of family labour

It can be seen that cost C is a very comprehensive concept. But the risk & uncertainty costs are not taken care of.

Variable Costs

These are the costs which are of the recurring type, & have to be incurred during every production period, e.g. seed fertilisers & insecticides. These are also known as operational costs or working costs.

Fixed costs. These costs are of non-recurring nature including cost of the tractor & other machinery buildings, irrigation structures, livestock, etc.

Ccrops costs. These costs refer to the value of seeds & plants, manures & fertilisers, insecticides & fungicides & irrigation charges.

Machinery costs. These costs include the hiring charges of machinery, cost of fuel & lubricants, electricity bills, minor machinery repairs, depreciation & interest on machines & equipment.

Labour costs. They include the wages paid in cash or kind to the hired labour & the imputed value of family labour used on the farm.

Livestock costs. They include the costs of veterinary medicines & services, fodders, feeds, interest on the values of livestock & depreciation.

Land costs. They refer to the rent of land owned & taken on lease.

Building costs. They include the interest on the value of the building structure & depreciation on them.

Sundry Farm Buildings

Farm Workshop Facilities

A workshop provides a focal point at the farmstead for the repair and maintenance of machines, implements and structures. It also provides a place where tools can be stored in an orderly manner, a store for supplies and spare parts, and a shelter where work can be carried out during inclement weather. A facility of this type should be available on every farm. The size and design of a workshop, however, should be commensurate with the size of the farm and the work to be done in the shop.

The small holder may be adequately served with a tool storage cupboard that can be locked for security and a workbench with a simple homemade vice for holding tools while they are being sharpened or fitted with new handles. From this simple beginning a more complete facility may gradually evolve as the farm operation grows and more equipment is required. Since repair tools and supplies represent a considerable investment, most farmers will want to store them in a secure place. Many small scale farmers will not require a separate store for this purpose, but if stored together with hand tools and small implements, the number of items may motivate the farmer to build a storeroom by enclosing part of the workshop with solid walls. A simple work shelter and store suitable for repair work and the storage of small implements. Note that the doors to the store may be designed with racks and hooks to hold supplies and tools. Fuels and other combustible materials should not be stored with the tools. A simple work bench and vice can also be housed under the shelter.

At the other extreme, a large ranch or commercial farm may need a separate building with extensive equipment for maintaining the farm machinery, tractors and vehicles. A farmer may also use his workshop to do routine repairs and preventive maintenance during the off season, to build or modify some of the equipment used on the farm and to prefabricate building elements to be used in construction projects.

The workshop facilities should be cost effective. That is, enough savings should be realized from timely maintenance, repairs and construction projects to pay for the cost of the building and the necessary tools and equipment. Although it is difficult to put a monetary value on timeliness, there is no question that being able to make emergency repairs is important. Some farm operations (planting, spraying, milking) are more sensitive than others to prolonged interruptions, and having facilities to complete repairs on the farm can reduce delays to a minimum.

Other factors, apart from the farm size, which will influence the extent of the workshop facilities are the number and diversity of machines, the availability of service from dealers, and the interest and mechanical skill exhibited by the farmer and farm labourers. If necessary, a skilled mechanic may be employed. Without qualified personnel to use the shop it becomes questionable in value and may even contribute to more frequent breakdowns and additional expense due to careless work. The workshop should be located close to the work centre of the farm and convenient to the farm home on ground that is well drained and sufficiently level to allow easy maneuvering of equipment. Where electric power is available, proximity to the power source should be considered.

In tropical climates the workshop may be a simple pole structure with a non-flammable roof. Unless dust is a problem, it may be feasible to leave the sides open to provide good light and ventilation. Heavy-gauge wire netting can be used to make the area more secure without reducing light or ventilation. A pole structure of this sort can be enclosed with offcuts or corrugated steel at a later time, but if this is done, there must be provision for several good-sized windows. While a simple earth floor is often satisfactory, concrete offers the advantages of an easily cleaned, level surface. To do a clean repair job, a clean work area is essential and this is particularly important when lubricated mechanisms are reassembled. The level surface is helpful in some assembly or alignment operations.

The following additional features are important for a safe and efficient shop:

- Sufficient room for the largest machine that may need repair, including workspace around it. If the machine is large, truss roof construction may be needed to provide the required space without intermediate supports.
- An entrance that is both wide enough and high enough for the largest equipment that the shop has been designed to accommodate. If the building is enclosed with either solid walls or wire netting, a second door is essential for safety in case of fire.
- Some means of lifting and supporting heavy loads. When the roof span is 3m or less, a timber beam is often adequate. For larger spans or very heavy loads a truss will be required. Alternatively, a portable hoist can be used.
- Electric lighting and electrical service for power tools.

- A water supply for both convenience and safety.
- One or more fire extinguishers of a type suitable for fuel fires. Two or three buckets of dry sand are a possible substitute or supplement for a fire extinguisher.
- Storage cabinets for tools, supplies and spare parts. Sturdy doors can be locked for security and also provide space to hang tools and display small supplies for easy access.
- A heavy workbench attached to the wall or otherwise firmly supported. It should be 1 m high, up to 800mm deep and at least 3m long and equipped with a large vice. There must be sufficient clear space around it to maneuver workpieces and, if attached to a solid wall, ample window openings above it to provide light.

Equipment needed in the workshop will depend on the type and extent of work to be done. Generally this means those tools required to perform day-to-day maintenance on machines and to carry out general repair work and small construction jobs required on farm buildings and equipment. However, any shop, regardless of size, will need some simple woodworking tools, some means of sharpening field tools, and wrenches (spanners) of various types and sizes. If the shop equipment includes a welder, it should be located, in the interest of safety, away from the woodworking area and preferably near the main door where it can conveniently be used inside or outside the building.

Flammable materials such as sawdust, shavings and oily rags must never be allowed to accumulate in the workshop since they represent a fire hazard, and fuels should be stored in a separate area. Generally good order and cleanliness in the shop makes for efficient work, convenience and safety.

Machinery and Implement Storage

On many small-scale farms in Africa all cultivation and transport operations on the term are done manually. The few small-sized hand tools and implements used in such farming can normally be stored in any multipurpose store at the farmstead. The store needs only to be secure for protection of the equipment from theft and vandalism, and dry so as to avoid deterioration of the metal and wooden parts. The tools will last longer if they are cleaned and working surfaces are greased prior to storage. The tools may be hung on rails or hooks on the wall or from the ceiling for order and convenience and to protect them from dampness penetrating an earth floor in the store. Implements such as ploughs, harrows and cultivators are damaged little by rust when left outdoors. If they are properly cleaned prior to storage and metal surfaces, particularly all threaded parts used for adjustments, are greased, then a little rust is not likely to harm performance enough to justify the cost of a storage structure. A fenced compound can offer adequate protection against theft during storage. Although implements containing wooden parts are more susceptible to decay, those parts can usually be replaced at low cost. Tractors and other complex machines will function better when needed if they have been stored under cover and given a complete off-season check-up. An adequate storage structure for these machines is likely to be economically feasible.

For most purposes a narrow open-side shed with a welldrained, raised earth or gravel floor will be adequate for machinery storage. The sides of the building can be partly or wholly enclosed with netting or solid walls when security conditions make it necessary. The building must be high enough to accommodate the highest machine. A smooth, level floor makes it easier to attach and detach tractor-mounted equipment or to move other machines. The space required can be determined by obtaining the dimensions of all the machines and implements to be stored. Then, using graph paper, the outline of the machines can be sketched onto a plan view, allowing additional space for maneuvering. Any roof-supporting posts inside the building or in the open sides must be marked on the drawing, since they will restrict the way the floor space can be utilized. Since many machines can not be easily moved, it is desirable to arrange the stored machines so that shifting is unnecessary.

Fire resistant construction is desirable where tractors, cars and other powered machines are stored. A pole structure with an earth floor, sheet metal walls, timber trusses and metal, asbestos-cement or sisal-cement roofing will offer adequate fire resistance. Machinery stores and farm workshops are constructed in much the same way and are usually placed close together for convenience. In fact, they may be housed in one building with a workshop section at one end and machinery and implement storage in the balance of the building.

Fuel and Chemical Storage

Many materials that are used on farms fall into the category of "hazardous materials", since they are either highly flammable or poisonous. The type and quantities of these materials requiring storage will vary from one farm or one cooperative store to the next and only a few basic requirements for safe storage will be considered here. Other materials frequently used on farms such as fertilizers and cement also have special storage requirements mainly because they are hydroscopic, i.e., they tend to pick up moisture from the atmosphere.

Storage of Hazardous Products

- Hazardous materials stored on farms normally include the following:
- Highly flammable materials such as engine fuels and oils, such as petrol, diesel, kerosene and lubricating oils.
- Gases such as butane, propane and acetylene. Oxygen promotes the combustion of other materials and must be handled carefully.
- Paints containing flammable solvents, cellulose thinner or alcohol.
- Poisonous materials such as herbicides, insecticides, rat poison and sheep and cattle dips.
- Acids and alkalies such as detergents, cleaning liquids, lye and quicklime (CaO).
- Medicines such as veterinary drugs and supplies. Some drugs may require refrigeration.
- Wood preservatives and corrosion inhibiting paints.

Hazardous materials should always be stored in a separate location containing only those materials. If the quantities are larger, flammable and poisonous materials should be stored in separate rooms. Ideally each type of material should have its own storage space, that is, its own shelf in a cupboard or a storage room, or its own room in a cooperative or merchant store. Quantities of flammable products greater than about 3 litres of cellulose thinner, 10 litres of petrol, 20 litres of kerosene, 50 litres of diesel fuel should be stored in a separate building at least 15m from any other building. For this purpose a pole building with steel netting walls offers shade and security. Any store for hazardous products must be well ventilated so that explosive or toxic fumes can not accumulate. Ventilation openings should be provided at both low and high levels or alternatively the door can be covered with netting. The store, including the ventilation openings, should be vermin proof to prevent rodents from breaking open packages. It must be possible to lock the store to prevent the theft of expensive materials and keep unauthorized persons, in particular children, from accidentally coming into contact with the hazardous materials.

Some chemicals are harmful to the skin. Therefore washing facilities should be available nearby for immediate use. Stores for hazardous materials should never have a drain in the floor as any spillage or washdown water containing the materials must be prevented from entering any watercourse or drinking water source. It is frequently recommended that the floor and lower part of the walls including the door sill be constructed of concrete to form a reservoir to contain any accidental spills. This type of store must be clearly marked with an appropriate warning notice.

Storage of Fertilizers and Other Non-hazardous Materials

Some fertilizers are hydroscopic and easily pick up moisture from humid air or from the ground. This causes them to become lumpy and to deteriorate. Cement, although not very hydroscopic, will deteriorate if exposed to damp conditions. Other materials may be adversely affected by prolonged exposure to high storage temperatures and therefore must be shaded. Fertilizers and cement are normally sold in plastic lined bags offering some degree of protection. They should be handled and stored so that the bags are not punctured or otherwise damaged. In addition the storage conditions should be as dry as possible. Bags should be placed on a raised platform in the store. This will allow ventilation and prevent ground moisture from penebating from below. The pile should be protected from rain by a roof or some other type of watertight cover. Fertilizer can be very corrosive to metals and should not be stored close to machinery or tools.

Greenhouses

A greenhouse is a structure using natural light within which optimum conditions may be achieved for the propagation and growing of horticultural crops, for plant research, or for isolating plants from disease or insects. While in the tropical areas of Africa there are only limited applications, there are a few situations in which a greenhouse can be justified because of the optimum growing conditions required for a high value crop or a research project. There is a wide range in the cost of various greenhouse designs and a careful assessment to relate the requirements for a given enterprise to the cost of the house is important. For example, a greenhouse used for year long flower production can justify the cost of glass, while a house used for a month or two for starting vegetable plants can only justify a polythene covering.

Site and Support Facilities

Greenhouses should be located in open areas with no shading from trees or buildings and with access to roads. The land should be nearly level and well drained with a fall of 1 in 100 to l in 200 being ideal. If possible, the site should be sheltered from excessive winds. However, normal air movement is essential for natural ventilation systems and to prevent locally stagnant conditions. Good soil is essential, deep, medium-textured loam being ideal. Soils which are less than ideal should be worth improving. Very heavy soils are not usually satisfactory.

A good, clean water supply is of paramount importance. A full crop system may require up to 8,400m^3 per hectare (840// m^2) in a single year and the source of water must be able to supply all that will be required. Electricity will be required if ventilation is to be mechanized and if stationary machinery is to be used in the greenhouse.

Design Parameters

Light

It is important that the crops being grown in a greenhouse receive the optimum amount of light, not only when the skies are clear (direct light), but also when it is cloudy (diffuse light). The shape and construction of the house should be such that it will allow the best possible entry of light. The two shapes coming closest to the ideal are: a the single-span semicircular section covered with clear polythene film, the mansard profile, a framed structure in which the sides and two roof sections are sloped in such a way that a semicircular cross section is approximated. The size and cross section of all the load bearing members have a pronounced effect on light transmission. The gutters of multi-span roofs produce considerable shade, and likewise, in wide-span houses, the heavier roof trusses tend to cause more shading. Thus open trusses with narrow-section members are desirable. Light colours and reflective surfaces improve light transmission. In spite of a good design for natural light, artificial lighting may be needed for the production of photo-period sensitive plants.

Orientation

Within the latitudes found in the tropics it is desirable to orient the ridge of greenhouses north and south to reduce the overall shading by the framing members. This is true for all types of frames including multi-span houses.

Size

While multi-span blocks of 3.2m each are least expensive to build, wider spans will allow somewhat better light transmission. Furthermore, the general management in wider houses (movement of machines, optimum cropping layouts, etc.) may justify the extra cost. As a general rule the cost is lowest when the length is four to five times the span width. This is particularly true with wide-span houses.

Height

The height of a greenhouse should be sufficient for the operation of machinery and the comfort of the workers. An increase in height improves natural ventilation during still conditions and the desired plant climate is more easily obtained. However, with very high roofs, maintenance becomes more difficult. Gutter heights of 2.8 to 3.0m are recommended for multi-span houses to allow machines to move freely. In single-span houses, eave height should be at least 2m to allow for unrestricted work space.

Materials

Greenhouses are generally built of steel, aluminium or wood and are glazed with good quality glass, clear polythene sheet, or fibreglass-reinforced polyester panels. Steel must be galvanized after fabrication as any welding or drilling breaks the galvanized layer. Steel is cheaper than aluminium and is ideal for the main roof frame. Aluminium is very resistant to corrosion and is easily formed into complex sections. While it is expensive, it is most suitable for glazing bars. It cannot be economically welded and bolted construction is used. Wood is less suitable for the lightweight construction and the high moisture conditions found in greenhouses, therefore only top grade timber of the most decayresistant species which has been treated with a waterbourne type of wood preservative should be used.

Glass is expensive, but it is the most durable covering and transmits the most light (90%). However, the gradual build-up of dirt and algae along with surface etching eventually causes a reduction in light transmission. The minimum width of glass ordinarily used is 610mm. Also common is the 730mm width. Both of these are 4mm thick and weigh 2.8 kg/m^2. Polythene sheet is increasingly being used to cover relatively low cost structures. It has light transmitting qualities similar to glass but the material has to be replaced periodically as it deteriorates under the influence of ultraviolet light. However, the cost is much lower than glass and the roof framing can be much lighter, resulting in good economy. Fibreglass reinforced polyester panels are more impact resistant than glass and more durable than polythene sheet. Light transmission is about 85% but drops off appreciably unless the surface is cleaned and resurfaced with acrylic sealer every 4 to 5 years. It is intermediate in cost between glass and polythene.

Ventilation

In tropical regions ventilation is likely to be the most important environmental control feature of the greenhouse. The exchange of air inside the building with air from the outside is used to lower temperature, reduce humidity, and to maintain a supply of carbon dioxide for photosynthesis. This is accomplished by natural means with vents and doors or by mechanical means with fans. A comprehensive discussion of ventilation is found in Chapter 7. The ventilation rate is usually expressed as the cubic metres per second of airflow per square metre of floor area.

To obtain a reasonable heat rise of less than 4°C in a glass-clad house, the airflow rate in the tropics should be 0.04 to 0.05m^3/s and m^2 of floor area. Polythene-clad houses do not become as hot due to the transparency of the plastic to longwave radiation which is transmitted back out of the house. Thus the ventilation rate for a polythene-clad house can be reduced to 0.03 to 0.04m^3/ s and m^2. This further reduces the cost of a polythene-covered house. Adequate natural ventilation is often provided by large doors at each end even though this may amount to only 3 to 7% of the floor area. These large doors not only aid in ventilation but also allow easy access to the greenhouse.

Cooling

Evaporative cooling can be used in greenhouses where ventilation alone is insufficient to maintain the required temperatures. The temperature reductions possible with evaporative cooling.

Shading

Shading is used to reduce light transmission and heat gain when necessary. In glass houses shading may be done simply by applying water-based whitewash to the inside of the roof to cut down light transmission. When the weather conditions are steady and reliable, whitewash is cheap and effective and easily washed off when the need is past. Whitewash as a shade seems particularly appropriate for shading in tropical areas.

Dairy Building Design and Construction

Site Selection

The following aspects should be considered when selecting a site for the dairy processing room:

- Water supply and quality;
- Milk supply "catchment";
- Land availability and quality;
- Other buildings and activities near the site;
- Proximity to a road:

- Effluent disposal; and
- Good drainage.

Water Supply

Water serves many functions in a dairy, such as washing, indirect heating, cooling milk and adjusting product composition. Water comes into direct contact with the product. It is important, therefore, to locate the dairy near a plentiful supply of clean water. Water can be collected from the roof of a dairy building or nearby barns by putting guttering around the roof and directing the water to a storage tank.

Land Availability and Quality

When selecting a site one should allow for possible future expansion. The site should be well drained.

Other Buildings

It is important to locate the dairy correctly in relation to other buildings. It should not be located near a hay barn or animal feed store where mould spores and dust are present as they can contaminate the raw material and products. It should also be located away from other sources of contamination such as dung heaps or cattle assembly areas.

Proximity to the Road

For convenience in collecting milk and for product distribution, the dairy should be located near a road. However, if the building is too near the road dust contamination will be a problem. Therefore, the doors and windows should not face the road. Windows for letting in light only can face the road.

Effluent Disposal

The satisfactory disposal of effluent from the dairy is important. Since most effluent comes from washing and from spillage, it can be minimised by careful product recovery, proper processing practice and care to avoid spillages. Rinsings and wash water should be piped away from the building for a distance of at least 15 meters and directed into a soak pit. Raw effluent should not be piped directly into a river or stream. If the effluent is not piped away from the building it will become a source of contamination and of foul smells.

Type of Building

A simple building of 25 m^2 internal floor area is adequate for processing at the scale being discussed. An additional room of 10^2 m floor area is desirable for use as a product store and office.

Construction Materials

The foundation and floor should be constructed from non-rotting material. The material for the superstructure is best chosen according to availability and cost. The dairy can be made from basic materials and does not need extravagant construction.

Floor

Where possible, all floors should be constructed of concrete with cement surfacing. The floor should slope (1–1.5%) to one end to facilitate drainage and cleaning. The cement should continue up the internal walls for at least one metre if the superstructure of the building is not constructed from cement.

Effluent Piping

The sloped floor drains to an outlet. Effluent should be piped from the outlet to the soak pit through concrete pipes 10 cm in diameter.

Light

One or two screened windows should be installed to permit the operation of the dairy without artificial light. The windows can also be used for ventilation, but should be screened with mesh to reduce the number of insects entering the building.

Ceiling

Where possible, a ceiling should be included. This will help to improve the hygiene of the building and also keep the inside cool.

Door

The main door should be wide enough to allow for equipment installation and easy access of personnel with milk cans etc.

Arrangement and Installation of Equipment

Arrangement

When arranging equipment one must consider:

- The flow of raw material through to product i.e. process sequence.
- Access to each item of equipment for operation, cleaning and storage.
- Storage and sale of product.
- Disposal of byproduct.

If both cheese and butter are being made, the process lines should be located at either side of the dairy.

Installation

Some items of equipment must be securely fixed. The cream separator should be mounted on a level stand fixed firmly to the floor and should be at a convenient height for working. Once the separator is mounted on the stand the level should be checked with a spirit level before final tightening of the fixing screws. Similarly, the churn stand and butter-working table should be fixed. Cheese vats of the necessary capacity are portable and can be located as desired. The fixing block for a lever-action cheese press should be fixed firmly to the wall. If the water supply permits it, two hose points should be installed on opposite walls to facilitate cleaning.

Chapter 6

Dairy Calf and Dairy Heifer Management

Cost of production is a key indicator of competitiveness. This is especially true in the dairy farm industry, where a manager must use input efficiency and resulting cost control to generate profit. There is no single method to calculate cost of production. Farm management consultants, university extension personnel, industry analysts, government economists, and farm managers all may have a unique, preferred method to calculate cost of production. The variation across farms with regard to enterprise mix, production technology, labour use, size, record systems, goals, and managerial ability contribute to the variation in cost of production methods. Tight margins and price uncertainty are among the reasons that cost of production is crucial in determining competitiveness and viability of farms.

Cost of production values are used for most major management decisions on the modern dairy farm. Perhaps the most widely used method to calculate cost of production involves using the farm income tax statement (Schedule F). Another method is enterprise accounting, which breaks out the milking herd and replacement heifer enterprises but allocates costs across all relevant enterprises on the farm. Comparing these methods reveals strengths and weaknesses.

In this paper we use detailed data from Michigan dairy farms to examine the cost of heifer raising. Examining the values generated by comprehensive enterprise accounting methods reveals several potential sources of errors, such as accrual adjustments, tax depreciation rather than economic depreciation, allocation errors, and unpaid factors including labour, management, and capital. Adjustments to modify the income tax method are considered to arrive at values acceptable for management decisions.

Cost of Heifer Raising

The Schedule F income tax form contains farm cash expenses and income for a calendar year. Because everyone has this information, it is a simple place to begin a cost of production calculation. However, one must use caution when using tax information to make management decisions. There are two major problems with the Schedule F method to calculate cost of production for an enterprise such as heifers. First, cash accounting is grossly inaccurate in capturing true costs because it records cost only when the money changes hands. This aspect is appropriately exploited by farmers to manage tax bills by shifting income and expenses across tax years. In addition, many costs are noncash (or even unpaid). The second issue is that this method makes it very difficult to allocate revenue and expenses to the appropriate enterprise. In most instances, the enterprise examined becomes the residual claimant of revenue and expenses. The result is that many adjustments must be made to arrive at a cost of production that is useful for management decisions.

There are a couple of basic fixes to the Schedule F method that can greatly improve its accuracy, depending on the characteristics of the operation in question. For example, accrual adjustments for inventory changes should be made to correct some of the cash basis flaws. Also, the value of unpaid labour and management can be estimated and included. Many operations have a computerized accounting system, and with some extra effort, this system can be tailored to deliver a much more usable cost of production. At Michigan State University we undertook a detailed enterprise accounting system of a set of dairy operations. By doing so, the intention was to provide an accurate picture of the relative strengths and weaknesses of enterprises on Michigan dairy farms and, also, to assess the appropriateness of simpler procedures to estimate cost of production.

Enterprise accounting involves setting up profit, cost, and service centres on the farm as well as allocation, or holding, accounts. Profit centres are enterprises that have both explicit revenue generation and costs. The milking herd, replacement heifers, and crops such as corn and hay are common profit centres for Michigan dairy farms. Cost and service centres are enterprises that exist to contribute to the profit centres. Often, the revenues or income for the cost centres is not explicitly paid; rather, they are a transfer within the same farm. Machinery, equipment, and buildings are examples of cost centres on dairy farms.

Allocation, or holding, accounts were set up to keep track of costs and revenues that apply across multiple profit or cost centres. If the cost or revenue could be readily assigned to a cost or profit centre, then that may be done during the standard accounting. However, there are many costs and revenues that apply across profit and cost centres and whose exact allocation is not known until later. An example of an allocation account is feed that is purchased for use by both the milking herd and replacement heifer enterprises. In the case of feed, the costs are allocated using the ration records. Other costs that require allocation include management, utilities, and fuel.

Data Recording and Collecting

The records needed to participate in this project were, for the most part, similar to what the farm cooperators were keeping with a few notable additions. Required information included:

1. *Financial Transactions:* The participants of the project utilized a computerized accounting program such as Microtel from MSU, Quicken, or QuickBooks. Entering financial transactions into the accounting system included assigning each transaction an enterprise code. This allowed tracking expenditures and income by profit centres or service centres.
2. *Beginning and Ending Inventories:* Balance sheets and the depreciation record had the market values used in valuing equipment, facilities, livestock, crops in storage, and other similar inventory items.
3. *Supplemental Physical Data Record:* This record was new. Its main purpose was to show how the inputs supplied by the service centres of the business are employed in the profit centres. For example, the cooperators recorded hours of labour as it was used by the corn or heifer enterprises. Information was recorded for the utilization of major machinery items in a similar way. This information was subsequently used to allocate the costs related to these service centres to the profit centres.
4. *Feed Utilization Sheet:* This record tracked the rations and the length of time they were fed to the various livestock groups. Often, the only extra effort involved was making a copy of rations being fed, noting the number of animals being fed the ration and the time period the rations were fed. From this information we allocated costs of both purchased and grown feeds to the livestock profit centres.
5. *Crop Production Record:* Most producers kept a log of crops harvested. This information was transferred to a summary sheet.

6. *Monthly Animal Inventories:* If the participant was on Dairy Herd Improvement Association (DHIA), this information was already recorded. If not, a barn sheet was supplied to record animal numbers by group (e.g., cow in milk, dry cows, heifers less than a year in age) on a monthly basis.

These records were checked for accuracy and later verified in summary form by the farm cooperators. Additional information such as the relevant market places for replacements, land rental, and opportunity cost of unpaid management and family labour was collected as needed.

Data Analysis

Costs were allocated from the holding accounts, such as feed to the appropriate enterprise. The crops that were harvested by a given farm were sold from that crop enterprise into a storage and marketing account. The milking herd and replacement heifer enterprises then purchased the feed from the storage and marketing account throughout the year at the going market rate. In this way, the storage and marketing account would gain any increase in market price since harvest, but would also be responsible for any loss and the cost of maintaining the storage facilities. Labour and management-supplied labour were allocated across enterprises based on the supplemental physical data forms. The value of labour was the going market rate for that quality of labour. A management fee was charged to each of the profit centres at 4 percent of the value added by the profit centre.

Owner capital was valued based on industry risk premium values estimated using MSU dairy farm industry data (Wolf, Hanson, Wittenberg, and Harsh). The individual farm charge for capital was estimated using the leverage position for the farm. The lowest cost of capital charge was 8.16 percent, while the highest cost of capital charge was 10.37 percent. An appropriate average value was 9 percent.

The total cost, which might also be referred to as the total economic cost of production for the enterprise in question, includes all factors of production, such as unpaid labour, management, and a competitive return to capital. This value did not represent a cash flow notion of costs that many producers are aware of; rather, it represented a complete accounting of all costs without any "free" use of unpaid resources, such as operator labour and capital.

Enterprise Accounting Results

To protect farm anonymity, all values are presented only in summary form, including average, minimum, and maximum. For each table, the minimum and maximum represent the extreme value for that production, revenue, or cost component across the farms. Note that neither the high nor low total costs are the sum of the individual costs by category. The replacement heifer enterprise was present in some form on all farms. The milk cow enterprise sold each heifer calf to the heifer enterprise as a newborn. The heifer enterprise raised the heifers and sold late pregnant (springing) heifers back into the milking herd. Each transaction was at current market price and validated by the producers.

Because the heifers were of all different ages and the farms were followed for a single year, an inventory adjustment was made, and all expenses are expressed as average cost per heifer month. That is, the costs and revenues were adjusted for the fact that older heifers were worth more than younger heifers at the end of the year. To put the comparisons on equal footing, we chose 24 months as the age to first calving to standardize the values. The range of heifer prices was $1,200 to $1,850 at first calving. This price reflected what the farm operators determined heifers were worth as they entered the milking herd. The standardized, 24-month-old, average heifer was valued at $1,300 (the last couple of years have witnessed heifer prices substantially higher than this).

The costs of production are divided into variable and fixed costs. The largest costs of raising heifers are the same as for the milking herd — feed and labour. The feed costs averaged $25.81 per heifer month (remember, these are across all ages of heifers), while labour averaged $11.06 per heifer month. Calf purchases includes purchasing the heifer calves from that farm's milk cow herd, while outside heifer purchases are heifers purchased from other operations. The "other" category includes various expenses such as repairs and utilities. The remaining expense categories are straightforward with total average variable cost equal to $54.37 per heifer month.

Fixed costs include depreciation on facilities, equipment, interest, insurance, taxes, and a charge for management. The fixed expenses averaged $13.289/heifer month for a total average cost of $67.65 per heifer month. Assuming a 24-month age to first calving, the average total cost to raise a heifer (with all factors, including unpaid expenses, included) for these farms was about $1,620. Note that this age assumption is perhaps a bold one because the age varied both within and across farms but was necessary for comparison across farms.

The minimum and maximum columns are across farms for that cost category. That is, the lowest average total cost to raise a heifer was $61.15 per heifer month, while the maximum cost was $67.65 per heifer month (these totals are not equal to the sum of each cost category, which are themselves minimums or maximums). Those farms that had lower costs in some categories often were higher in other cost categories.

Even with the range in farm size and location, the farms examined were remarkably consistent in almost every cost category. The net average loss per heifer across these farms was $21.23 per heifer month, or about $480 per heifer. This reflects the difference between what the producers thought the heifer was worth as a springing heifer and what it cost to raise the heifer to 24 months. This value applies only to these farms and is not necessarily representative. It is possible that the producers did not assign a large enough value to the springing heifers. The numbers do, however, reflect all the costs involved for all factors of production, and the key point is the methodology.

Implications for Cost of Production Estimates

Cost of production is a key performance indicator in the dairy industry. Heifer raising enterprises can use a cost of production estimate to make decisions. The most accurate cost of production method is accrual enterprise accounting, which allocates all costs. It is probably impractical to expect all producers to track all the required information. However, there are several key actions that can make cost of raising heifer estimates more accurate.

First, be certain to use accrual accounting because cash methods ignore inventory changes, which are very important in agriculture (consider how many expenses you prepay in December of a good income year).

Second, unpaid labour, management, and capital should be included in the cost estimate. Labour and management should be valued at their opportunity cost. That is, if you were to hire labour, what would it cost (including any relevant benefits)? Similarly, what would replacing the unpaid management functions cost? The capital investment should also earn a return — it could be in stocks or bonds, for example. We found that 9 percent was a good estimate for the risk level on Michigan dairy farms.

Finally, allocating overhead, feed, fuel, and other expenses that are spread across multiple enterprises is very important if the farm has more than one major enterprise. We had farm managers track labour and machinery uses in notepads. We found that if they kept careful track for a week or so during each season, the allocation was fairly stable. This means that you need not track your efforts for months at a time. But take the time to track a week and use that as a basic rule of thumb to allocate unpaid labour and management. The same implication applies to allocating machinery and equipment that is used for multiple enterprises.

Heifer Economics

Today's successful dairy operation recognizes that heifers are an important investment in the future. They place high value on the heifer and regard it as a managed resource, whether raised on the farm or contract grown. Unfortunately, on many farms, the dairy heifer is the most overlooked and under managed asset on the farm. The main goal for managing replacement heifers is to freshen them between 22 and 24 months of age to reduce expenditures and to increase total milk production. This can be accomplished through good nutrition and sound animal management practices. Other areas of focus include genetic improvements. This can be made through a well planned and thought out breeding program. Maintaining a good health program during the heifer-raising period is important. This includes vaccinations, deworming, and any necessary treatments. The success of a heifer-raising program is directly related to its overall economic management. The purpose of this paper will focus on the economic awareness needed for developing a long-term heifer raising enterprise.

Economic Awareness

The costs involved in raising heifers should be an important issue for dairy farmers. Replacement animals typically account for 15 to 20 percent of milk production costs. Replacement heifers rank as the second or third largest component of production costs after feed and possibly labour on most dairy farms. These costs can vary from farm to farm depending on individual management strategies. The cost of raising heifers is influenced by two main concerns, management and economic. The management concerns are:

1. Herd morbidity and mortality rates.
2. Age at first calving and herd replacement rates.

The economic concerns are:

1. Ownership costs
2. Operating cost

Management Concerns

Herd Morbidity and Mortality Rates

Minimizing calf morbidity and mortality rates involves a combination of management components from a good dry cow vaccination program, colostrum management, sanitation, and proper nutrition and care of the newborn. In addition, it includes a variety of preventative measures as well as maintaining good health practices. Ultimately, controlling calf health problems will save many times the cost of these practices with reduced heifer raising costs.

Age at First Calving

Age at calving and herd replacement rates are the largest factors influencing heifer costs. This affects the numbers of heifers that must be raised to maintain a profitable milking herd size.

Table 1 summarizes the numbers of heifers that must be maintained at various levels of herd replacement rates and ages at first calving. These two factors alone have a major impact on the overall costs.

Table 1. Heifer herd size for a 100-cow herd and a 10% heifer cull rate.

Cull Rate (%)	***Age at First Calving (months)***				
	22	***24***	***26***	***28***	***30***
26	53	58	63	67	72
30	61	66	72	78	83
34	69	76	82	88	94
38	77	84	92	99	106
42	86	93	101	109	117

When age at calving increases, so does the need for heifer housing, feed, labour, and management. This increase in input variables can be as much as 50 % or more in extreme situations. An example of this magnitude of increase would be comparing a farm with a 26 % herd turnover rate and 22 month calving age with another having a 38 % herd turnover rate and a 30 month calving age. The first farm would need 53 heifers in the replacement herd, while the second would require 106, or twice as many heifers just to maintain a constant 100 cow herd size. The costs to raise these extra heifers can be tremendous and make a major difference in the profit potential of each farm.

Economic Concerns

It is important for farmers to understand the total costs involved in raising dairy heifers. In order to operate a successful enterprise it is necessary to know the current costs in order to predict future costs.

Ownership Costs

Ownership costs include buildings, equipment, property, machinery, depreciation, interest on investment, repairs, taxes, and insurance. Many of these things that may seem obvious to the owner may get overlooked. Care should be given to include all ownership costs when evaluating a heifer-raising program. Each ownership cost adds significantly to the overall cost of raising a heifer.

Operating Costs

Operating costs include feed, labour, bedding, utilities, veterinary care, breeding costs, and supplies. These vary nearly proportional to the number of heifers raised at one time. With good record keeping most variable costs are easily understood and calculated.

Table 2 gives a typical breakdown of heifer expenses, both ownership and operating, from birth to prefreshening. The table shows the results from a spreadsheet that calculates the costs to raise a replacement heifer. Aspects that need to be included when calculating the cost to raise a heifer are feed, labour, breeding, bedding, health, buildings, equipment, mortality, and interest.

In Table 2, costs are separated by the following age periods:

- birth to weaning
- weaning to 6 months
- 6 months to first bred
- bred to prefreshening

Calculating costs in this manner provides managers the ability to evaluate areas of strength and weakness within their heifer-raising program. Feed costs usually constitute 60% of the total overall expense to raise heifers. The most expensive age period in feed cost per heifer is birth to weaning. This is due to the large labour and feed costs per animal.

Labour costs calculate the time required raising a heifer. Every aspect has some cost associated with it, in this instance, the cost of time. Labour costs are the second highest expense in raising a heifer, around 13% of the total cost. Breeding costs includes both artificial insemination and the use of a service bull. The use of a service bull is not a cheap breeding source. Maintaining a service bull on a heifer operation incurs cost such as the interest on the purchase of the bull, feed, and labour to manage the service bull. These costs are often quite large on a per heifer basis. Bedding, health costs, buildings, and equipment costs are all necessary to calculate the costs to raise heifers.

Even though buildings and equipment may have depreciated their value to $0, these items still require maintenance that must be calculated as real costs.

Table: *Costs of raising replacement dairy heifers.*

	Birth until	***Weaning until***	***6 mo until***	***Bred until***	***Totals***
	Weaning	***6 mo***	***1st Bred***	***Prefresh***	
Operating Cost					
Feed	$50.82	$112.46	$228.02	$309.83	$701.18
Labour	$20.87	$60.00	$30.24	$71.27	$182.37
Breeding				$21.60	$21.60
Bedding	$2.56	$6.42	$1.35	$11.02	$21.35
Health		$7.90	$1.62	$2.81	$12.33
Capital Ownership Cost					
Buildings	$0.91	$3.64	$20.86	$34.61	$60.02
Equipment	$0.90	$3.54	$6.93	$6.18	$17.55
Animal Ownership Cost					
Mortality	$0.60				$0.60
Interest	$1.05	$6.80	$15.36	$48.33	$71.54
Miscellaneous	$10.00	$10.00	$10.00	$10.00	$40.00
Totals	$87.71	$210.75	$314.43	$515.65	$1,128.54
Age at Weaning/d	42	Age at 1st Bred/mo.	15	Age at Pre-. Fresh/mo	23.5
Per Day	$2.25	$1.53	$1.17	$2.02	$1.60
ADG/lbs	1.5	1.8	1.8	2.0	1.87
Per lb/gain	$1.50	$0.85	$0.65	$1.01	$0.86

Mortality and interest costs calculate the opportunity cost of raising a heifer. Mortality cost is associated with the loss of the investment, while interest cost is the opportunity cost of having capital invested in a heifer versus the bank. These two cost estimates are the most overlooked items when calculating heifer-raising costs. This is because opportunity is a non-tangible product that cannot be seen or touched. However, it is a cost that must be calculated. Mortality and interest costs constitute the third largest expense. It is impossible to estimate every cost that contributes to the total cost to raise a heifer. Expenses that are difficult to estimate include water, power, fuel for equipment, and time to transport or move heifers. Thus, when calculating the cost to raise a heifer it is advisable to incorporate a miscellaneous cost figure that can cover these costs.

Ownership costs for the heifer operation are often hard to calculate and easy to overlook. It is important for farmers and heifer growers to know what the actual cost of their heifer-raising program is before changes can be made. Economic awareness will be the most valuable tool the farmer has for making sound operating decisions. In addition, by knowing their actual heifer-raising costs, some farmers find that the alternative of having someone else raise their heifers (contract grower) is a cheaper alternative.

Specialization in the dairy industry and pressure to make sound economic and environmental decisions has created a need to evaluate each dairy management decision. Research has shown that the two most expensive age periods based on total cost per day to raise a heifer, are birth to weaning and bred to prefresh. Selecting management methods that can decrease the length or expense of these periods can have a significant impact on the total cost to raise a heifer.

Other Areas of Improvement

There are some aspects of calf and heifer raising that can be more efficient than others. Many farms can benefit by reducing some cost components in their replacement program without reduction in heifer quality. The following is a listing of some areas that should be considered as potential cost saving areas for heifer raising on most farms.

1. Feed a lower cost source of liquid feed to young calves.
2. Feed high quality and palatable concentrates to younger animals.
3. Analyze forages and run ration formulations for all major groups.
4. Monitor group size and age/weight variation within groups.
5. Use proven feed additives to improve growth and feed efficiency.
6. Keep weight gains steady at 1.8 pounds per day before nine months of age and 2.0+ pounds per day after nine months of age.

Feed a lower cost source of liquid feed to young calves.: Depending on a variety of aspects available to the farmer, changing from whole milk to other liquid feeds can be cost effective. Milk replacers are often about 50 60 percent of the cost of feeding whole milk, if salable milk is fed. Where farms are set up for feeding waste milk and colostrum in a safe and easy manner, this can be even more cost effective. Waste milk systems are not without problems or increases in management, however many farms can handle the additional problems that arise. Waste milk and colostrum must be fed in a consistent manner to avoid health problems, and if possible, pasteurized to minimize any health problems and disease transfer.

Feed high quality and palatable concentrates to younger animals.: This means feeding the best quality calf starters and calf growers to the young calves. A high quality starter, one with no mold, no dust, with a good texture, high levels of nutrients, and plenty of molasses and or flavouring agents, will make a dramatic impact on early starter and dry matter intake. Optimizing starter intake will allow calves to grow at higher rates of gain and will allow for earlier weaning ages. Typically, calves need 1.5 to 2 pounds of grain per day for at least three days prior to weaning. Once weaned, calves require significantly less labour and if they continue to grow at rapid rates, will be much more economical per pound of gain then when fed liquid diets.

Analyze forages and run ration formulations for all groups.: Since forages make up a large part of heifer diets, they must be sampled and analyzed for nutrient content in order to achieve balanced diets for these animals. Slightly or severely misbalance diets will not be utilized nearly as well as those that are balanced. When forages make up a large part of the diet, even a small difference between estimated and actual analysis will be costly. In addition, allocating forages to the various age groups that will best utilize them is an aspect of forage feeding that can be cost effective.

For example, high protein forages should be fed to younger aged heifers that have a high protein requirement. Lower protein forages should be fed to older aged heifers with lower protein requirements and increased gut capacity for less nutrient dense forages.

Monitor group size and age/weight spreads within groups.: This aspect of heifer feeding management means monitoring the different ages and sizes of heifers within a group, and how well their nutrient needs are being met by the diet for that group. Smaller heifers will tend to under-consume and larger heifers will over-consume rations if grouped together and fed a restricted diet.

Use proven feed additives to improve growth and feed efficiency.: The use of ionophores has proven effective in improving feed efficiency and/or growth rates of heifers. This improvement is in the range of 5 to 7 percent or more, and is well documented in the scientific literature. These compounds also have other benefits for the dairy heifer including control of coccidiosis. The cost/benefit ratio is extremely favourable in using these compounds.

Keep weight gains steady.: Often farmers get heifers too fat or too large early in life and cause these animals to require more nutrients to be needed for maintenance later in life when their growth rates are less. Unless rapid early growth is used with earlier calving, this practice is not as efficient as raising heifers at a steady rate of gain as needed for the desired age and weight at calving. Growing heifers too slowly in early life is also expensive as it requires more nutrients in later stages of heifer development, increases age at calving, or reduces body weight at calving. All of these are detrimental to overall heifer economics.

Research has consistently demon-strated that rate of gain greater than 1.8 pounds per day before nine months of age will cause a decrease in first lactation milk production. After nine months of age, increasing growth rate to 2 pounds per day or more will achieve larger size heifers at calving. Heifer raising costs can be great if they are not controlled and evaluated given the farm and the required end product. Periodic monitoring of heifer economics will pay great returns for dairy farmers.

Heifer Management Blueprints: Heifers and Feed Bunk Management

One simple way to improve feed efficiency in heifers is to employ good bunk management. Feeding heifers to exact levels of intake and using the heifers' inherent nature to sort feed as a guide to manage bunks has been demonstrated to improve feed efficiency. Paying proper attention to eating behaviour and managing the feed bunk accordingly can increase feed efficiency and decrease feed cost.

Start with a Good Bunk Design

Feed is meant to be consumed by animals. A properly designed feed bunk for heifers should first and foremost minimize feed losses behind the feed bunk. Research data from Michigan State University demonstrated that up to 20 percent of feed can be lost to the aft side (behind) the feed bunk. In general, feed losses will be less when heifers are required to place their heads through and reach down for feed, as opposed to reaching horizontally for feed. Feed wagons where the feed is located at the same horizontal plane as the animal's muzzle have been demonstrated to increase feed losses.

Fence line feed bunks should be properly fitted for each size group of heifers. Post and rails, throat guards, and/or self-locks should be checked and adjusted to proper dimensions. Listed in Table 1 are minimum bunk space requirements and suggested dimensions for post-and-rail feeding fences and waters.

Use Feed Sorting as a Management Tool

Research at the University of Wisconsin has demonstrated that heifers will sort feed similarly to lactating dairy cows. Heifers, as with lactating dairy cows, will choose to consume the shortest particles first and refuse long feed particles. Because long forage particles and/or corncobs generally contain more NDF or less energy than small feed particles such as grain, heifers may consume diets higher in energy than formulated. Likewise, if heifers are allowed to refuse long feed particles, they will not reach fill limitations as soon and, subsequently, dry matter intake will increase.

The effects of reducing feed offerings to heifers, forcing the heifers to consume all or most long feed particles. The result is a slight decrease in feed intake, which results in saved feed and improved feed efficiency.

Feed Adjustments in Small Increments

Research data from South Dakota State University suggest heifers (or steers) should not be overfed on a daily basis. Precisely monitoring and controlling feed intakes and feeding the heifers to exact intakes (very minimal feed waste) will reduce feed wastage and increase feed efficiency. The combination of proper bunk design and feeding heifers to exact intakes has been shown to result in 10 to 15 percent improvements in feed efficiency. To feed heifers to exact intakes, a bunk scoring vocabulary should be utilized. A simplified bunk scoring vocabulary is:

Score	*Definition*
0	No feed remaining
1	A few small scatter particles of feed remaining
2	Many feed particles remaining, concrete still visible
3	Large amounts of feed remaining, no bunk concrete visible

The objective of a controlled bunk management feeding system is to feed heifers to a bunk score of 1 every day. If bunks are empty (Score 0) or excessive feed is remaining (Scores 2 and 3), then feed intakes are moved up or down in small increments (2 percent) to facilitate feeding heifers to a bunk score of 1. This type of feeding system also helps assure heifers consume all large feed particles and feeds such as corncobs.

Full consumption of diet also assures the formulated diet is actually being totally consumed. Bunk design and management are often overlooked in heifer management programs. Proper bunk management has been demonstrated to increase feed efficiency 10 to 15 percent, which directly results in a 10 to 15 percent reduction in feed cost or in an annual feed cost savings of $30 to $40 per heifer.

Table: *Minimum bunk space requirements and suggested dimensions for post-and-rail feeding fences and waters.*

	3-4 Mo.	*5-8 Mo.*	*9-12 Mo.*	*13-15 Mo.*	*16-24 Mo.*
Feed Always Available					
Hay or Silage	4 in./head	4 in./head	5 in./head	6 in./head	6 in./head
Mixed Ration or Grain	12 in./head	12 in./head	12 in./head	18 in./head	18 in./head
All Animals Eat at Once					
Hay, Silage, or Ration	12 in./head	18 in./head	22 in./head	26 in./head	26 in./head
Throat (height, in.)		14	16	17	19
Neck Rail (height, in.)		28	30	34	41
Maximum Water (height, in.)	29	31	33	34	

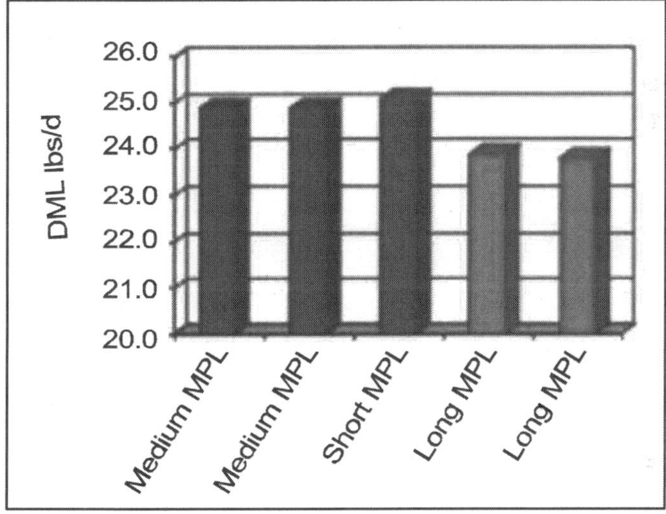

Figure 1.

Heifer Management Blueprints

Dairy producers and heifer growers often house dairy heifers in free-stall barns as a labour-efficient heifer management system. There are many unique aspects to consider when managing heifers with a free-stall housing system.

The Benefits of Free-Stall Housing

The use of free-stall barns restricts the resting environment for heifers, allowing for easier observation, bedding, and manure management. Primarily, free-stall housing systems allow for efficient manure management and reduction of bedding. In comparison to a bedded pack, free-stall barns save on space needed per animal. The use of headlocks in free-stall housing systems also allows catching heifers with relative ease to facilitate vaccinations, breeding, and other routine management tasks associated with heifer management.

Negative Aspects of Free-Stall Housing

One of the biggest drawbacks of free-stall housing systems is that free-stall resting surfaces do not work well for heifers less than 400 pounds in cold climates. This is due to a difference in heifer biology as compared to lactating cow biology. Young heifers weighing less than 400 pounds have more surface area (square inches of hide) per pound of body weight than lactating cows. As a result, when small heifers rest on something cold, such as cement, a mattress, or sand, net energy is lost through conduction. Thus, the free-stall resting surface may be too cold for a lightweight dairy heifer to rest on and maintain energy balance for proper growth. Heifers less than 400 pounds housed in free-stall barns often look energy deficient during cold weather, with abnormally rough hair coats and abdominal distention. Heifers under cold stress may also be more prone to respiratory disease.

Quirks of Managing Heifers in Free Stalls

Listed below are several quirks associated with managing heifers in free-stall barns:

- Heifers are difficult to move in free-stall housing systems. They are not adjusted to moving laterally down the alleys. Thus, management tasks such as weighing heifers on a scale may be challenging because heifers are not accustomed to walking up and down the alleys to remote areas of the barn. Because heifers are not adjusted to moving laterally down the manure alleys, drover's alleys better facilitate moving heifers to remote locations (hoof trimming, scales, breeding chutes, etc.). The addition of drover's alleys will result in more management flexibility but increase the cost of the facility. Therefore, headlocks are most often used as a method of animal restraint in heifer free-stall barns as compared to post and rail feeding systems.
- Managing heifer foot health can be a problem in free-stall barns, especially hairy heel warts. Because heifers do not move well laterally down the alleys, moving heifers to a common footbath or hoof-trimming stall can be challenging. Multiple or movable footbaths in the cross alleys may be required to manage foot health issues.
- Free-stall housing systems present a quandry when it comes to removing manure off the alley surface. Because heifer free-stall barns often contain multiple groups of heifers, skid-steer

scraping systems require opening and closing of multiple gates. In contrast, mechanical alley scrapers are more prone to freezing in heifer free-stall barns than cow free-stall barns because of the type of manure produced and the number of animal units.

- Free-stall floor surfaces are often slippery, which can result in some compromises in mounting behaviour associated with heat detection, etc.

What Type of Bedding Should be Used?

The manure handling systems will most likely dictate the bedding used in a free-stall barn. The primary types of free-stall bedding are mattresses and sand. Using sand as bedding has been demonstrated to provide more comfortable resting surface for the heifer, but bedding usage and labour will be higher as compared to mattresses. Sand-laden manure may require sand settling lanes or mechanical sand separation systems. Mattress free-stall systems are typically bedded with a layer of sawdust and result in a more flexible manure handling system, but they require bedding at more frequent intervals.

Bunk Space and Overcrowding

Overcrowding of a free-stall facility may help optimize the economics of the facility, but it may result in more heifers lying in alleyways, which are wet concrete surfaces not suitable for resting heifers. Trauma injuries can also result from heifers being housed in free-stall barns too small for their body size. Most notably, swollen hocks and other trauma injuries have been reported from the use of small stalls. On the other hand, large stalls allow for urine and feces to fall in the resting area of the animal. This can lead to more labour time spent cleaning stall surfaces, as well as a higher rate of mastitis in heifers. While research is limited, a general guideline is to not overcrowd heifer free-stall housing systems more than 125 percent.

Access to proper bunk space may be as much or more of a critical reason to avoid overcrowding than concerns with resting area. First, limited bunk space might disallow any opportunity to limit-feed heifers, which has been demonstrated to significantly improve feed efficiency and decrease manure production. Limit-feeding strategies cannot be implemented when bunk space is inadequate. Second, severely limiting bunk space may result in shift feeding, feed sorting, and, ultimately, uneven rates of average daily gains.

Managing Heifers in Free-Stall Barns is Different

Every heifer housing system is a compromise between cost, labour efficiency, and animal environment. Each of these should be taken into consideration when choosing a housing system.

If you have a stable herd size and even calving distribution, free-stall barns can be a labour efficient and economically sound way to house animals greater than 400 pounds. If the choice to use a free-stall housing system for heifers is made, you must then choose what type of free-stall operation you would like to operate. The manure management systems available will most likely dictate bedding, manure scraping systems, and, ultimately, barn configuration. Other critical management systems to consider are feeding, hoof health, and animal restraint. You must also ensure your free stalls are of adequate size for the animal. They must allow adequate space for lying down as well as lunge space and appropriate neck rail and brisket board height to minimize injuries. Recommended free-stall dimensions are listed in Table below.

Table: *Recommended free-stall dimensions for dairy heifers.*

Age (months)	*6-9*	*9-12*	*12-15*	*15-19*	*19-24*
Weight (pounds)	330-560	560-710	710-860	860-1050	1050-1290
Free-stall Width (inches)	30	33	37	42	45
Side Lunge Free-Stall Length (inches)	60	64	72	78	81
Forward Lunge Free-Stall Length (inches)	72-78	76-82	84-90	90-96	93-99
Neck Rail Height (inches)	31	35	38	41	43
Curb-to-Neck Rail and Brisket Board (inches)	46	49	57	62	64

Labour Management for Heifer Rearing

Good labour management is a key to the success of the modern dairy business. Labour management includes the delegation of tasks, the supervision of employees, and the completion of tasks in an orderly and timely manner. In addition, employees should be aware of the consequences when there are lapses in management and when tasks are done incorrectly. When a manager delegates responsibilities to employees, it implies that the manager knows their skills and knows the standard operating procedures (SOP) for the particular job. SOPs are important to maintaining a safe working environment as well as optimizing animal performance. It is unrealistic to expect an employee who does not possess the necessary skills or is poorly trained to do a job correctly.

Performing a task properly can be defined by following the proper SOP for the job being conducted. SOPs are written procedures on how a particular job is to be completed. Having this information in writing minimizes miscommunications between the employer, manager, and employees. There are four basic points to keep in mind when managing labour and ensuring that SOPs are being followed. The first step is to meet with the employees and discuss how current protocols are being implemented. This can help to determine if employees are following the SOPs, if they have been trained properly or if the SOPs need to be updated.

A second point for SOPs that deal with everyday tasks is to list events step by step. It is often the small details that are assumed to be known which create the major problems. A third recommendation is that SOPs be written in the most simplistic language so there are no misunderstandings on how, when, and where tasks should be preformed. Lastly, all employees should be aware of the SOPs and they should be posted in highly visible areas. This helps new employees as well as part-time help to perform tasks correctly.

SOPs should allow tasks to be completed in a timely and consistant manner everyday. This can assist the manager in making efficient use of all resources available, not only labour.

Chore Chart for a Labour Management System

To ensure that managers and employees alike understand their duties and responsibilities it is a good idea to develop a chore assignment or accountability system. A simple chore chart idea has been developed to use with each of the four logical dairy heifer-growing periods. This chart is designed as an example of one farm only. Its intention is to help employers and managers think of new and better ways to enhance the current labour management system. However, it should be noted that an attempt was made to model this after a typical Pennsylvania dairy farm that is raising replacement heifers. In the spreadsheets, the frequency of the chores is specified. The other factors specified will vary and depend on the unique situation of each farm business. The outcomes and consequences are also given as examples and will vary across individual businesses.

Chore

Chores will be self-explanatory in most cases. The personnel and general management chore for the owner or manager is to set aside time for managing and training employees. It should also include an allowance for the owner or manager to undertake training or self-education workshops as well. Examples of continuing education include attending extension or industry workshops and farm visits. Time should be allotted for reading journals, newspapers, magazines, or conducting Internet searches for information to improve the overall efficiency of the dairy business.

Daily, weekly or monthly: This is the number of occasions per time period that a particular chore should be done.

Personnel: The person who should do a particular job or is primarily responsible for the task. A well-trained employee is one who the owner or manager believes can do the task satisfactorily without direct supervision or very limited supervision.

Time: An approximation of the time to do the task specified.

Labour cost: The cost of the time taken to do the task specified, based on the time taken and the cost of the labour of the person doing the task.

Responsibility factor (RF): (1 lowest – 5 highest). The responsibility factor is a measure of the importance of doing the task correctly and in a timely manner. A high responsibility factor implies that the consequences of not doing the task correctly can be extremely costly to the dairy farmer.

Outcome: The result of completing the task correctly or at the right time.

Consequence: The result of carrying out the job incorrectly or not at the right time.

Summary of Chore Charts

Calves from birth to weaning: Daily tasks related to feeding are the most frequently required tasks for this age group, however the tasks of feed preparation and equipment sanitation have the highest responsibility factor. They require a well-trained employee or the owner or manager handles these tasks. These items also have the highest potential consequence if they are done improperly. Health checks are also high priority tasks and have great conse-quences in both time and economic loss. The other tasks are important, yet not as critical to the day-to-day health and well being of the calves. The other tasks must be done occasionally as noted in order to maintain a controlled calf raising operation. The degree of record keeping and personnel or general management will vary greatly with the size of operation and individual goals of the farm owner. Therefore, the time related to these tasks are not given.

Heifers from weaning age to 6 months: In this age group, proper feed preparation, regular health checks, and personnel or general management are the most important factors to accomplish. Other than feeding and repairs (when necessary), the other factors related to this age of heifers are more flexible as to when they are accomplished. Thus these animals can serve as a time buffer for some of the rest of the farm operation. Record keeping however is of high relative importance and is critical to the success of the farm enterprise and animal well being. Part of this record keeping is to monitor the growth of the animals. This should include weight (scale or heart girth tape) and height of the heifer (wither height or hip height). These monitors encompass a variety of items, but they are most directly related to feed quality, ration balance, and general health program.

Heifers from 6 months of age to breeding: Tasks, relative importance, followed by outcomes and consequences are similar to the previous age group. Some of the daily tasks are of slightly less relative importance as the immediate nature of the consequences is not as great. For example, if a ration is not accurately prepared on one day, the consequence is trivial. If not prepared correctly for a week, it could be important. Long term poor animal husbandry practices are of great consequence and management is still of top relative importance. During this time period heat detection and breeding become factors of top importance. Time spent involved in these aspects can have a tremendous impact on overall heifer rearing costs relative to herd average age at calving.

Heifers from breeding age to prefreshening: Similar to the previous heifer group, as heifers get older and larger, the day-to-day relative importance is less critical to the outcome, however chronic weekly or monthly problems are important. Management must have some degree of quality control to maintain proper and consistent growth rates in a cost-effective manner.

Making a custom chore chart: Some tasks and chores can be done in conjunction with each other, providing the person doing the chore has the training. An example of this is the person responsible for feeding calves can also check the health of these animals as the feeding is being done. The calf feeder may need some training to be able to identify animals that may be unwell or symptoms that indicate an animal is unwell and can report to the person responsible for health checks.

Improving Dairy Heifer Reproductive Management

Reproductive management of dairy heifers is certainly one area in which many dairy producers can improve. Heifers are the most fertile animals in a herd and should have the greatest genetic potential. Dairy producers need to pay more attention to getting heifers bred sooner and breeding them to genetically superior bulls. Heifers should calve at 24 months. This is 2 months lower than the Georgia average. Heifers should be bred artificially to top A.I. sires to further increase the genetic potential and minimize calving difficulties. Currently, a lot of heifers are pasture bred to genetically inferior herd bulls and most producers raise their own herd replacements.

Puberty and Estrus Activity

Puberty in dairy heifers is more related to body weight than age. Heifers reach puberty at 30 to 40 percent of the average mature weight. At this time, the hormonal patterns that regulate the estrus cycles begin developing and result in the heifer coming into heat on a regular basis. The first few heats may be erratic and anovulatory (no ovulation). After a couple of cycles, heifers should cycle every 20-21 days (range 18-24 days). Puberty may be delayed when under-feeding, diseases or parasites delay growth. Feed heifers a balanced ration based on their age and weight. Protein, energy, minerals and vitamins can affect reproductive function. Your veterinarian should be consulted for the most efficient worming and vaccination schedule to use on heifers. Poor weight gains, lower conception rates, poor body condition, rough hair coats and elevated body temperatures have also been associated with cattle that graze fescue pastures infested with fungus. Growth of dairy heifers on fescue pasture can be affected by the level of fungus infestation. Most fescue pastures have a level of fungus that can affect conception rates. Record heat cycles prior to the breeding season, and have a veterinarian examine heifers that exhibit abnormal cycles or are anestrus and not detected in heat. Heifers should be at the proper weight and size for breeding at 15 months of age. This will allow heifers to calve when they are 2 years old. Research has shown that heifers that calve after 27 months produce slightly more lifetime milk. However, the higher costs to carry these heifers outweigh any increase in milk yield. Keep average daily gains below 2.2 pounds per day to avoid excess fat. Research has shown that this increase in fat can show up in the udder and lower production. Producers should breed heifers around 15 months using the following table:

	Body Weight (lbs.)	*Heart Girth (in.)*
Jersey	500-600	58-60
Ayshire & Guernsey	650-700	61-63
Swiss & Holstein	750-800	64-66

The only sure sign a heifer is in heat is standing to be mounted by another animal. Animals are only in heat 8 to 15 hours. Therefore, it is important to check heats twice daily for 15 to 20 minutes each. It is also important to watch for secondary signs of heat. These include a clear mucous discharge, mounting other animals, a red swollen vulva, nervousness, bawling, dirty flanks and head-laying on other animals. Blood found on the tail is not abnormal. Animals showing this sign were in heat one to two days ago and should be watched more closely in 17 to 19 days.

Successful reproductive management operations generally have several people in charge of both heat detection and artificial insemination. Someone must be responsible for getting the job of heat detection done accurately on a regular basis. All heifers should be clearly identified. Heifers should be checked twice daily for 20 minutes each time to see over 70 percent of the heat activity. Twenty-one day records should be kept on all cycling heifers.

Heat mount detectors, marking crayons or tail paint, teaser animals and prostaglandins, can be considered as aids to help a producer detect more heats. Also, training heifers on self catching gates will make injections and heat detection aid applications easier as well as help them be ready to adjust better at calving.

The HeatWatch* estrus detection system (DDX, Inc., Denver, Colorado) was the first detection device with 24-hour surveillance potential for standing activity. Components of the HeatWatch system are a miniaturized pressure sensitive radio transmitter powered by a lithium 3 volt battery and secured in a water resistant pouch attached to a nylon mesh patch; signal receiver (1 mile range); buffer which stored activity data until downloaded to software that enables routinely generated management lists. Activation of the sensor sends a radio telemetric signal containing code for transmitter identification, date, time, and duration of standing event.

When and where to watch for heat is also important. Animals on dirt or pasture are more active than on concrete. During hot weather, heifers should be watched late in the evening, after the sun goes down. However, during winter, it is best to watch during the middle of the day when temperatures are generally the warmest. Moving animals from one location to another also increases activity.

Artificial Insemination

Ovulation occurs approximately 30 hours after onset of heat. Highest conception rates are achieved if a heifer is bred 12 to 14 hours after onset of heat. With good heat detection, satisfactory conception rates can be obtained by breeding according to the A.M./P.M. rule. If a heifer is observed in heat in the morning, inseminate her in the afternoon. If a heifer is observed in heat in the afternoon, breed her the next morning. Once a day breeding can also be used if heifers are bred in the morning. Heifers must, however, be checked for heat twice daily.

Using top A.I. bulls on all heifers gives a producer an efficient and economical way to increase milk production. Results will show up two years down the road in the milk parlour. Daughters of A.I. sires have more than a 1,200-pound milk production advantage over non-A.I. sires.

Select a group of high net merit bulls to use on a group of heifers. Select those bulls from a specific group that produce the fewest difficult births or have shown to produce the fewest difficult births. Select bulls that are the best buys both economically and genetically. Pasture breeding with natural service may be convenient and possibly efficient, but should not be used because of genetic reasons. Also, bulls are dangerous, and hurt or kill people each year as well as carry diseases and can possibly be subfertile. Cull infertile breeding heifers of ample size as they generally do not get any better. After breeding, a veterinarian should check animals for pregnancy by 60 days.

Controlled Internal Drug Release

Calving Management

Move heifers into the calving area two weeks prior to calving. Provide shade to reduce stress. Like for cows, give grain and gradually increase it to allow the rumen time to adjust prior to calving. Observe heifers close to calving at frequent intervals. Like cows, heifers should be given a clean, dry place to calve. A grassy pasture or bedded maternity pen works best. Calving assistance should be given if the heifers fail to calve within 2 hours after onset of labour. Disinfect hands and equipment, and ask for help from your veterinarian when you need it.

Conclusion

Use the following guidelines in the reproductive management of heifers:

- Heifers should be grown to attain adequate size for breeding at 15 months of age.
- Watch heifers for heat activity twice daily for a minimum of 20 minutes each.
- Breed heifers to superior A.I. bulls known to produce few difficult births.
- Use estrus synchronization procedures to breed heifers.
- Heifers can be bred once daily in the morning.
- Remember that heifers are the most fertile animals in your herd and generally have the greatest genetic potential.

Chapter 7

Judging Livestock

Livestock judging consists of carefully analyzing animals and measuring them against a standard that is commonly accepted as being ideal. Livestock judging also has been defined as a study of the relationship between an animal's form and function.

There are numerous benefits to gain from competing on a livestock evaluation team. In the course of training and competition you are given the opportunity to interact with future leaders of the livestock industry. You develop a keen sense of judgment and confidence to make a decision that you defend in a set of oral reasons. Most prominent livestock people who are masters of judgment and selection have been affected by their involvement with livestock judging. Participation in livestock judging builds your character and makes you a more complete person. "Judging instills the confidence in those people who may be timid and humbles those who tend to be conceited." — Harlan Ritchie

Steps to Successful Livestock Judging

Techniques for livestock judging can be broken into four steps:

Information: You must develop a mental image of the ideal for the species, breed and sex involved.

Observation: Successful livestock judging requires a sharp eye and a keen mind. As a livestock judge you must learn to develop a greater perception for the animals that are being judged. Furthermore, these observations must be accurate and complete in every way.

Comparison: A successful judge must make comparisons of each animal against each of the other animals in the class. Judges must be able to weigh the good and bad characteristics of each animal and make their decisions based upon their findings. The animal that will eventually be selected to top the class will be the one with the "most of the best."

Decision: The final, and perhaps most difficult, step involves ranking, or placing, the animals in the class. Many people have the necessary information and the ability to observe and compare but they may lack the skill and courage required to make the right decision. Judges' decision must be accurate and logical, and they must stand by them.

Tips for competitive livestock judging:

- Do not waste valuable time. Begin judging as soon as the class is in the ring.
- Every class is different. Do not "read" anything into the class! Do not play hunches! Do not try to out-guess the official judges. Place the livestock exactly as you have been taught.
- Break each class down into sections: Top pair, easy bottom, close middle, etc.
- When the contest starts, do your own work. Depend entirely upon your own judgment. Pay attention to the class and not to anyone else. When the contest is over, discuss the classes with your instructor and the officials.
- Try to have a tentative placing before too much time has elapsed. Otherwise you may find yourself out of time with no decision made.
- When evaluating a class, your first impression is usually your most unbiased and most accurate if it is the result of careful analysis.
- For general observation of a class, maintain a minimum distance of 25 feet. This will allow you a full unobstructed view of the class. If other students move in too close, just ask your group leader to please move everybody back.
- When finishing a placing on a class, make sure to mark your card. Double check to make sure the placing on your notes matches the one on the card you turn in.
- If there is a break in the contest do not let the time go to waste. Make use of this by looking over your notes from previous classes.
- Nearly all students go into a slump at some time during their judging careers. When this happens, do not get discouraged with yourself or upset with your instructor. Just keep your wits, maintain a positive attitude and work hard — you will pull out of it.

Benefits of giving Reasons

There are two parts to the format when judging livestock: Placing the livestock and giving oral reasons on your placing. The second part can be the most difficult because you have to convince an official, who has already reached a decision, that your placing is logical whether the official agrees or not. This can be difficult and stressful, so what do you get in return from learning proper reason-giving?

- You learn how to become organized; you get your thoughts in line with what you have on paper.
- You learn to recall a situation clearly in your mind that happened several hours ago or longer.
- You learn to express yourself in a convincing manner.
- You become able to clearly enunciate words.
- You learn to defend your decisions. This not only benefits you in judging livestock, it is also an asset for everyday life.
- You learn how to sell yourself and your ideas.

Evaluating a Set of Reasons

There are several schools of thought concerning oral reasons. The style you choose to express yourself is of little importance. The truly important factors involved in giving an effective set of oral reasons include:

- Accuracy of statements.
- Completeness of all the important points.
- Ability to bring out the important points between pairs of animals.
- Complete vocabulary of livestock terminology.
- Term variation.
- Correct grammar.
- A sincere, emphatic and precise presentation.
- An appropriate voice level, which depends on the size of the room.
- Clear enunciation.
- Voice inflection.
- A logical order to your reasons.
- Proper presence, e.g., eye contact, correct posture, no distracting mannerisms.

- Use of animal identification, e.g., red barrow.
- Do not overuse industry terms, e.g., "more useful to the cattle industry."
- Use carcass terms in market classes.
- Ability to communicate from a visual image of the animals.

Preparing Notes for a Contest

Unless you are gifted with an unusual memory, good note taking is a must on reason classes. A stenographer spiral notebook (6 by 8 inch) is a good size to use. Remember that notes are used to help you visualize the animals in the class. Use your notes to refresh your memory of the animals. Avoid memorizing your notes; you should give reasons from a mental image of the animals rather than memorizing the notes.

Helpful hints on note taking:

* Before the contest, take your note pad and divide a page into the sections.
* When you approach a class, stand far enough away from the animals for a general look.
- Write down the big things that appear to you first: Size, thickness, volume, etc.
- Write down IDs of the animals — if not color, then shape.
- Write down a tentative placing.
- Start with your first pair and write down the assets for the top animal. Then write the grants for the animal you placed second. Proceed to your middle and bottom pairs.
- Do not spend the entire class time writing. If you do, you will memorize your notes and be unable to recall the animals in the class when you get ready to give your reasons.
- Your notes should be brief. Develop a method of shorthand to quicken the note-taking process. Above all, make sure your notes are readable.
- Do not waste time. If you finish a class early, relax, look at the animals and try to remember them. This will enable you to more easily visualize the class for reasons.

Proper Etiquette of Oral Reasons

Once you have developed a proper reason style, it is important to present them properly to the official. When you walk into the reasons room, the way you present yourself may be almost as important as what you have to tell the official. This is why proper etiquette is of utmost importance when delivering a set of reasons. Important things to keep in mind when you get ready to give a set of reasons include:

- Be prompt. When it is your turn to give a set of reasons, do not keep the official waiting. If you find yourself being rushed, just ask the official to extend your time. He or she may or may not.
- Always leave notes outside the reason room.
- Enter the reason room with an air of confidence but not cockiness.
- Some officials will offer you your card. If so, take the card, thank the official, place the card behind your back and continue your reasons. Do not depend on this card as not all officials will offer it to you.
- Do not stand too close to the official; 6 to 9 feet is about right depending on your size and your voice strength.
- Stand with your feet spread to about the width of your shoulders. Keep both hands behind your back.
- Stand erect. Avoid leaning over too far.
- Look the official squarely in the eye, or at least give that impression. Above all, do not let your eyes wander.
- Talk in a strong voice, slightly louder than a conversational voice, but do not shout.
- Speak with the utmost conviction and sincerity.
- Do not let yourself talk too rapidly as the official may not catch everything you say.
- Vary your delivery — make your main points impressive and emphatic.
- You may gesture slightly with your head; any other body gestures are too distracting to the listener.
- Avoid mixing numbers.
- Never exceed two minutes on a set of reasons. This is a rule for most collegiate judging contests.

Reason Format

The ability to give effective reasons is an important quality for a good livestock judge. Many factors influence the effectiveness of your reasons. However, unless reasons are presented in a manner that is pleasant to hear and clear and easy to follow, the value of accuracy is largely lost because much that is said does not "get through" to the listener. By following the traditional format used at MU, you can organize your reasons to cover all the points that were found in the class as well as keep the reasons short enough to remain in the two-minute time limit.

In the following format, we discuss the placing of crossbred market steers. Each pair is broken into three subsets:

- General statements
- Reinforcements
- Grants.

This format allows you to talk about the pairs in a logical order, which makes giving the reasons easier as well as making listening to them easier.

General statement: The most important factors for placing the pair this way. (Example: "In my top pair of red steers, I placed 1 over 2 because 1 is a thicker-made, larger-framed steer.")

Reinforcements: Go into more detail to back up the general statement. (Example: "1 is wider through his chest floor, wrapped more muscle around his forearm, is thicker down his top. He showed more skeletal height at the apex of his shoulder. 1 would appear to hang a carcass with more retail product than 2.")

Grants: Grant the biggest assets first. (Example: "I will grant that 2 is a trimmer steer and should display a carcass with a superior Yield grade.")

Middle Pair

General statement: "In a logical middle pair, I placed 2 over 3 as 2 ties together a more desirable combination of muscling and trimness than 3."

Reinforcements: "2 appears to push less condition through his shoulder pocket and has less visible fat through his flank and cod areas. 2 also shows more natural muscle volume down his top and through his quarter than 3."

Grants: "I will concede that 3 is more structurally correct than 2 as he has a more functional set to his rear legs."

Bottom Pair

General statement: "In my bottom pair and a close decision, I used 3 over 4 as he appears to be more able to reach the Choice grade.

Reinforcements: "3 has more visible fat cover down his top and over his ribs than 4. 3 also shows more natural thickness through all portions of his body and would appear to hang a more merchandisable carcass than 4."

Grants: "I will grant 4 is a trimmer designed steer, but 4 is the narrowest, lightest-muscled steer in the class. 4 has the least amount of packer appeal and therefore merits no higher placing in this class today."

When you are finishing with your last animal, be sure to finish strong with a closing statement so the official will know that you are through.

Reason styles: There are three basic styles of reasons used at MU:

Style 1: The more traditional style just discussed gives admissions last within a pair. This style has been used to allow our students to become more comfortable with giving reasons.

Example: "I placed this class of Hampshire Boars 1-2-3-4. In my top pair, I used 1 over 2 as he is a heavier weight, higher performing, more mobile boar. 1 exhibits more natural width through his chest, down his top and through his ham than 2. I will concede that 2 is a trimmer boar as evidenced by less fat deposited through his shoulder pocket, down his top and through his ham seam. 2 also stands squarer on his front legs."

Style 2: The second style used is an alteration of the traditional format. In this style, you present the criticisms of the top animal at the beginning of the first pair. We suggest this style for the more polished reason giver since it deals with different types of transitions that may be unfamiliar to the inexperienced individual.

Example: Introduce class: "I aligned this class of heavy structured Hampshire Boars 1-2-3."

Top Pair:

General statement: "Even though 1 is a splay-fronted boar, he easily excels 2 as he is a heavier weight, higher performing, more mobile boar."

Reinforcements: "1 exhibits more natural width through his chest, down his top and through his ham than 2."

Grants: "I will grant that 2 is a trimmer boar as evidenced by less fat through his shoulder pocket and through his ham seam."

Style 3: The third style is a combination of the previous two. It incorporates the criticisms of the animal in the opening sentences but leads off with a positive statement.

Example: Introduce class: "I liked the Suffolk Breeding Ewes 1-2-3-4."

Top Pair:

General statement: "I started this class with 1 as she exhibits the most superior combination of structural correctness and extension in the class. If I could complete the class winner."

Reinforcements: "Still, I used her over 2 in my top decision as 1 is a more nicely balanced, straighter-lined and a more eye appealing ewe that is leveler down her top and straighter and squarer on her rear legs. Furthermore, 1 is the most extreme about her growth indicators as she exhibits the greatest length of head, neck and cannon and the most potential growth and outcome in this class."

Grants: "I will certainly admit that 2 is the thickest-made, widest-based ewe in the class. She, too, is bolder sprung through her rib and exhibits more internal volume than 1. Yet, I preferred her in second as she is heavy fronted, coarse-shouldered and short-necked."

Middle Pair:

General statement: "Coming to the middle pair, it's still 2 over 3 as there is simply more ewe in 2."

Reinforcements: "2 is a larger framed ewe that is longer-bodied and taller made. Also, she appears pounds heavier being wider-based and thicker made than 3. In addition, 2 is a more capacious ewe that is deeper and bolder sprung and exhibits more width down her top and more dimension to her rack."

Grants: "I will grant that 3 shows more Suffolk breed character, being blacker about her points with a longer, more bell-shaped, ear. But I placed her third today as she is tight-ribbed, shallow made and fine-boned.

Bottom Pair:

General statement: "In closing, I still liked 3 over 4 as 3 is a more stylish appearing and angular ewe being more feminine about her head and especially smoother in her shoulder."

Reinforcements: "In addition, she projects more extension than 4 as she is especially longer about her head and neck. 3 also stands on more length of cannon and is a later maturing ewe than 4.

Grants: "I recognize that the blue tagged ewe is thicker-made and higher-volumed as she stands wider-based than 3. But that doesn't compensate for the fact that 4 is the smallest-framed, shortest-bodied and most conventionally designed. She, too, is a heavy-fronted, coarse-shouldered ewe that stands on the least length of cannon, and exhibits the earliest maturity pattern in this class."

With the second and third reason styles, do not let repetition slip into your format. They will be quite effective if used properly.

Words and phrases to avoid;

- Better — this word is too weak; it explains nothing.
- Animal or individual: say what the animal is (barrow, gilt, heifer, etc.).
- Lacks or lacking — non-descriptive; instead of saying a gilt "is lacking width," say "she is narrow."
- Beware of words ending with "ing." These words tend to be weak: placing, criticizing, faulting. Instead, say "I placed, I fault, I criticized," etc.
- Number — don't say "number 1 is;" instead say "1 is."
- Avoid excessive use of "he" or "she". Be more specific; use an ID; e.g., "the black heifer."
- Do not use the phrases "for being" or "kind of." Example: "I placed 3 last for being light muscled;" instead say "I placed 3 last because he is light muscled."
- Do not use "it;" every animal has a gender.
- Do not use the word "that;" e.g., "that rump," "that top," instead say "squarer-rumped" or "leveler-topped."

Young cattle producers will be faced with selection decisions that affect their profitability. They should use all information available, including performance data. Evaluation through performance records and visual appraisal better prepares students for realistic selection decisions. A cattle producer using performance information is like any successful business owner who uses the most accurate inputs possible to make economically sound decisions. Judging decisions are always controversial. The goal is to make a sound, defensible decision based on fact and to learn from the judging exercise how to improve cattle production through selection.

Beef Cattle Production

Beef is the culinary name for meat from bovines, especially domestic cattle. Beef can be harvested from cows, bulls, heifers or steers. It is one of the principal meats used in the cuisine of the Middle East, Pakistan, Australia, Argentina, Europe and the United States, and is also important in Africa, parts of East Asia, and Southeast Asia. Beef is considered a taboo food in some cultures, especially in Indian culture, and thence is eschewed by Hindus and Jains; it is also discouraged among some Buddhists.

Beef muscle meat can be cut into steak, roasts or short ribs. Some cuts are processed (corned beef or beef jerky), and trimmings, usually mixed with meat from older, leaner cattle, are ground, minced or used in sausages. The blood is used in some varieties of blood sausage. Other parts that are eaten include the oxtail, tongue, tripe from the reticulum or rumen, glands (particularly the pancreas and thymus, referred to as sweetbread), the heart, the brain (although forbidden where there is a danger of bovine spongiform encephalopathy, BSE), the liver, the kidneys, and the tender testicles of the bull (known in the US as *calf fries*, *prairie oysters*, or *Rocky Mountain oysters*). Some intestines are cooked and eaten as-is, but are more often cleaned and used as natural sausage casings. The lungs and the udder are considered unfit for human consumption in the US. The bones are used for making beef stock.

Beef from steers and heifers is equivalent, except for steers having slightly less fat and more muscle, all treatments being equal. Depending on economics, the number of heifers kept for breeding varies. Older animals are used for beef when they are past their reproductive prime. The meat from older cows and bulls is usually tougher, so it is frequently used for mince (UK)/ground beef (US). Cattle raised for beef may be allowed to roam free on grasslands, or may be confined at some stage in pens as part of a large feeding operation called a feedlot (or concentrated animal feeding operation), where they are usually fed a ration of grain, protein, roughage and a vitamin/mineral preblend.

In absolute numbers, the United States, the European Union, Brazil, and the People's Republic of China are the world's four largest consumers of beef. On a per capita basis, Argentines eat the most beef at 64.6 kg annually per person; people in the US eat 40.2 kg annually, while those in the EU eat 16.9 kg.

The world's largest exporters of beef are Brazil, Australia, and the United States. Beef production is also important to the economies of Argentina, Mexico, New Zealand, Nicaragua, Russia, and Uruguay.

Beef production begins with a cow-calf producer who maintains a breeding herd of cows that raise calves every year. When a calf is born, it weighs 60 to100 pounds. Beef calves are weaned at six to 10 months of age when they weigh in the range of 450 to700 pounds.

Calves normally leave their ranch or farm of origin between 6 and 12 months of age. Younger or lighter weight calves may be sent to a backgrounder or stocker who continues to graze them on grass or other forages until they are 12 to 16 months old. Both the cow-calf and stocker segments graze cattle on range and pastureland that is largely unsuitable for crop production. In fact, about 85 percent of U.S. grazing lands is unsuitable for producing crops and utilizing this land for beef production is an efficient use that more than doubles the area that can be used to produce food.

After the calves are weaned, some are sold at an auction market. A cow-calf producer may also choose to keep the best females to add to the breeding herd. Some animals may not be sold at an auction market, and instead will go directly from the cow-calf producer to the feedlot or from the backgrounder/stocker to the feedlot for the final growing phase in beef production.

Most beef cattle spend approximately 4 to 6 months in a feedlot just prior to harvest where they are fed a grain-based diet. At the feedlot (also called feedyard), cattle are grouped into pens that provide space for socializing and exercise. They receive feed rations that are balanced by a professional nutritionist. Feedlots employ a consulting veterinarian, and feedlot employees monitor the cattle's health and well-being daily. Feedlots are efficient and provide consistent, wholesome and affordable beef using fewer resources. The time cattle spend in a feedlot is often called the "finishing phase." Beef production through the feedlot system is safe and economical.

Some producers may choose to finish cattle on grass pasture. The beef derived from these animals is "grass-finished" and may also be referred to as "grass fed." Beef production from grass finished animals is said to appeal to certain healthy diets. This is a significantly smaller segment of modern beef production because it requires unique climate conditions, and it takes the cattle longer to reach market weight. All cattle, whether they are grass-finished or finished in a feedlot spend the majority of their lives grazing on grass pasture.

Once cattle reach market weight, typically 1,200 to 1,400 pounds and 18 to 22 months of age are sent to a processing facility to be harvested. U.S. Department of Agriculture (USDA) inspectors are stationed in all federally inspected packing plants and oversee the implementation of safety, quality and animal welfare standards from the time animals enter the plant until the final beef production phase ends and the products are shipped to retail and food service establishments for consumers to purchase.

You will need to study your beef cattle Herd Health and learn to spot an obviously sick animal and then diagnose what cause it to get sick in the first place.

Quite often an acutely sick animal will have been quite ill before the signs were noticed by the stockman. This cause other animals in the herd to be fighting off the same health problem and may be causing their to production levels to be going downhill. This is why you will need to develop an experienced eye towards detecting disease within your beef cattle herd. The quicker it is spotted and treated the better off you and your her will be.

Even though you are not expected to have the skills of a veterarinan you as an individual owner or stockman in charge of the herd must still accept full responsibility for your herd health program. This means that you alone must possess or quickly develop the skills of observation and management ability for the first line of defense to ward off any major health disasters. It almost goes without saying that you can afford a considerable amount of time to develop these skills. You will profit from it.

One major factor in herd health is the ability to identify each animal individually, this is fundamental to any herd health program. Each animal should be marked by ear tags, brand or some other visible identification method. The next step to follow would be good records to be kept based upon these identities. This will allow you to develop a herd health history on each animal and occasionally previous health issues may warrant culling. Without a herd health history you could easily miss this part. A good practice is to set up a planned program to help protect the herd health. Some of the items might be vaccination, castration, isolation, sanitation and herd replacements done on a systemized organized basis. Each replacement might be quarantined away from the rest of the herd for a given period of time before putting them together.

For sure it will be necessary to control or eliminate disease before it becomes a major problem. This will require the use of all the tools mentioned above in the fight for your herd health and at the same time remembering they are tools to be used at the right time and in the right place.

You will need to know the capabilities and limitations of animal drugs and learn how to administer them properly. Be sure to follow the labeled use restrictions unless instructed otherwise by you veterarinan.

You will need the proper equipment and facilities for handling and treating your beef cattle. This will reduce time and labour requirements to administer a herd health program and both you and the herd will benefit.

We are also told that the number one cause for most disease problems or bad herd health is a lack of nutrition. We must do our best to that our animals are fed at the level needed for maximum performance of the job expected from them. And last but not least, remember you veterarinan. Use him when needed but use hime wisely and economically.

Beef Cattle

Beef cattle are cattle raised for meat production (as distinguished from dairy cattle, used for milk production). The meat of cattle is known as beef. When raised in a feedlot cattle are known as feeder cattle. While the principal use of beef cattle is meat production, other uses include leather, and products used in shampoo and cosmetics.

Beef Cattle Breeds

Breeds known as *dual purpose breeds* are also used for beef production. These breeds have been selected for two purposes at once, such as for both beef and dairy production, or both beef and draught. Dual-purpose breeds include the Brown Swiss and many of the Zebu breeds of India such as Tharparkar and Ongole. The original Shorthorn was also a dual-purpose breed but diverged into two groups through selective breeding. A steer that weighs 1,000 lb (450 kg) when alive will make a carcase weighing about 615 lb (280 kg), once the blood, head, feet, skin, offal and guts have been removed. The carcase will then be hung in a cold room for between one and four weeks, during which time it loses some weight as water dries from the meat. When boned and cut by a butcher or packing house this carcase would then make about 430 lb (200 kg) of beef.

Value-Based Beef Cattle Production

Today's beef industry is evolving toward a concept of value-based marketing which prices cattle and carcasses on individual merit rather than averages.

Recognizing that consumer's wants and needs drive beef demand, those who contribute to the added value of superior products should be rewarded. Likewise, the market should penalize those responsible for producing an inferior product. Whether you are feeding a 4-H steer, an FFA beef project or are a commercial beef producer, this publication is designed to help you learn about the concepts of value-based beef cattle production and marketing. It consists of information you can use to help you take market beef animals from weaning (preconditioning) to finish. The major focus centers around producing a high quality beef end product for the consumer while adding profitability to the beef cattle enterprise.

Value-Based Beef Cattle Production Goals

I. Determine frame scores, beginning weights, desired end weights and calculations needed for average-daily-gains
   1. Measure hip-height and weigh beef cattle
   2. Calculate frame score, desired end weight and estimated average daily gain needed to reach desired end weight.

II. Building rations to meet goals
   1. Develop rations utilizing locally grown feeds to meet the goals for average daily gain and end weight set above.

III. Learn about carcass quality characteristics — what are they, how are they measured and how do they influence value on a live-weight basis?
   1. Develop an understanding of carcass quality characteristics, how they are measured and how they relate to both carcass value and live animal value.
   2. Develop an awareness of Beef Quality Assurance techniques and issues.

Chapter 8

Horse and Mule Production

The horse (*Equus ferus caballus*) is a hooved (ungulate) mammal, a subspecies of the family Equidae. The horse has evolved over the past 45 to 55 million years from a small multi-toed creature into the large, single-toed animal of today. Humans began to domesticate horses around 4000 BC, and their domestication is believed to have been widespread by 3000 BC. Although most horses today are domesticated, there are still endangered populations of the Przewalski's Horse, the only remaining true wild horse, as well as more common populations of feral horses which live in the wild but are descended from domesticated ancestors. There is an extensive, specialized vocabulary used to describe equine-related concepts, covering everything from anatomy to life stages, size, colors, markings, breeds, locomotion, and behaviour.

Horses' anatomy enables them to make use of speed to escape predators and they have a well-developed sense of balance and a strong fight-or-flight instinct. Related to this need to flee from predators in the wild is an unusual trait: horses are able to sleep both standing up and lying down. Female horses, called mares, carry their young for approximately 11 months, and a young horse, called a foal, can stand and run shortly following birth. Most domesticated horses begin training under saddle or in harness between the ages of two and four. They reach full adult development by age five, and have an average lifespan of between 25 and 30 years.

Horse breeds are loosely divided into three categories based on general temperament: spirited "hot bloods" with speed and endurance; "cold bloods", such as draft horses and some ponies, suitable for slow, heavy work; and "warmbloods", developed from crosses between hot bloods and cold bloods, often focusing on creating breeds for specific riding purposes, particularly in Europe. There are over 300 breeds of horses in the world today, developed for many different uses.

Horses and humans interact in a wide variety of sport competitions and non-competitive recreational pursuits, as well as in working activities such as police work, agriculture, entertainment, and therapy. Horses were historically used in warfare, from which a wide variety of riding and driving techniques developed, using many different styles of equipment and methods of control. Many products are derived from horses, including meat, milk, hide, hair, bone, and pharmaceuticals extracted from the urine of pregnant mares. Humans provide domesticated horses with food, water and shelter, as well as attention from specialists such as veterinarians and farriers.

Horse anatomy is described by a large number of specific terms, as illustrated by the chart to the right. Specific terms also describe various ages, colors and breeds.

Lifespan and Life Stages

Depending on breed, management and environment, the domestic horse today has a life expectancy of 25 to 30 years. It is uncommon, but a few animals live into their 40s and, occasionally, beyond. The oldest verifiable record was "Old Billy", a 19th-century horse that lived to the age of 62. In modern times, Sugar Puff, who had been listed in the Guinness Book of World Records as the world's oldest living pony, died in 2007, aged 56.

Regardless of a horse's actual birth date, for most competition purposes an animal is considered a year older on January 1 of each year in the northern hemisphere and August 1 in the southern hemisphere. The exception is in endurance riding, where the minimum age to compete is based on the animal's calendar age. A very rough estimate of a horse's age can be made from looking at its teeth.

The following terminology is used to describe horses of various ages:

- Foal: a horse of either sex less than one year old. A nursing foal is sometimes called a *suckling* and a foal that has been weaned is called a *weanling*. Most domesticated foals are weaned at 5 to 7 months of age, although foals can be weaned at 4 months with no adverse effects.
- Yearling: a horse of either sex that is between one and two years old.
- Colt: a male horse under the age of four. A common terminology error is to call any young horse a "colt", when the term actually only refers to young male horses.

- Filly: a female horse under the age of four.
- Mare: a female horse four years old and older.
- Stallion: a non-castrated male horse four years old and older. Some people, particularly in the UK, refer to a stallion as a "horse".
- Gelding: a castrated male horse of any age.

In horse racing, these definitions may differ: For example, in the British Isles, Thoroughbred horse racing defines colts and fillies as less than five years old. However, for Australian Thoroughbred racing, colts and fillies are less than four years old.

Size and Measurement

The height of horses is measured at the highest point of the withers, where the neck meets the back. This point was chosen as it is a stable point of the anatomy, unlike the head or neck, which move up and down. The English-speaking world measures the height of horses in hands (abbreviated "h" or "hh", for "hands high") and inches. One hand is equal to 101.6 millimetres (4 in). The height is expressed as the number of full hands, followed by a decimal point, then the number of additional inches. Thus, a horse described as "15.2 h" is 15 hands (60 inches (152.4 cm)) plus 2 inches (5.1 cm), for a total of 62 inches (157.5 cm) in height.

The size of horses varies by breed, but also is influenced by nutrition. Light riding horses usually range in height from 14 to 16 hands (56 to 64 inches, 142 to 163 cm) and can weigh from 380 to 550 kilograms (840 to 1,200 lb). Larger riding horses usually start at about 15.2 hands (62 inches, 157 cm) and often are as tall as 17 hands (68 inches, 173 cm), weighing from 500 to 600 kilograms (1,100 to 1,300 lb). Heavy or draft horses are usually at least 16 to 18 hands (64 to 72 inches, 163 to 183 cm) high and can weigh from about 700 to 1,000 kilograms (1,500 to 2,200 lb).

The largest horse in recorded history was probably a Shire horse named Mammoth, who was born in 1848. He stood 21.2½ hands high (86.5 in/220 cm), and his peak weight was estimated at 1,500 kilograms (3,300 lb). The current record holder for the world's smallest horse is Thumbelina, a fully mature miniature horse affected by dwarfism. She is 17 inches (43 cm) tall and weighs 57 pounds (26 kg).

Ponies

The general rule for height between a horse and a pony at maturity is 14.2 hands (58 inches, 147 cm). An animal 14.2 h or over is usually considered to be a horse and one less than 14.2 h a pony. However, there are many exceptions to the general rule. In Australia, ponies measure under 14 hands (56 inches, 142 cm). The International Federation for Equestrian Sports, which uses metric measurements, defines the cutoff between horses and ponies at 148 centimetres (58.27 in) (just over 14.2 h) without shoes and 149 centimetres (58.66 in) (just over 14.2½ h) with shoes. Some breeds which typically produce individuals both under and over 14.2 h consider all animals of that breed to be horses regardless of their height. Conversely, some pony breeds may have features in common with horses, and individual animals may occasionally mature at over 14.2 h, but are still considered to be ponies.

The distinction between a horse and pony is not simply a difference in height, but other aspects of *phenotype* or appearance, such as conformation and temperament. Ponies often exhibit thicker manes, tails, and overall coat. They also have proportionally shorter legs, wider barrels, heavier bone, shorter and thicker necks, and short heads with broad foreheads. They may have calmer temperaments than horses and also a high level of equine intelligence that may or may not be used to cooperate with human handlers. In fact, small size, by itself, is sometimes not a factor at all. While the Shetland pony stands on average 10 hands (40 inches, 102 cm), the Falabella and other miniature horses, which can be no taller than 30 inches (76 cm), the size of a medium-sized dog, are classified by their respective registries as very small horses rather than as ponies.

Colours and Markings

Horses exhibit a diverse array of coat colors and distinctive markings, described with a specialized vocabulary. Often, a horse is classified first by its coat color, before breed or sex. Horses of the same color may be distinguished from one another by white markings, which, along with various spotting patterns, are inherited separately from coat color.

Many genes that create horse coat colors have been identified, although research continues to further identify factors that result in specific traits. One of the first genetic relationships to be understood was that between recessive "red" (chestnut) and dominant "black", which is controlled by the "red factor" or extension gene. Additional alleles control spotting, graying, suppression or dilution of color, and other effects that create the dozens of possible coat colors found in horses.

Chestnut, bay, and black are the basic equine coat colors. These colors are modified by at least ten other genes to create all other colors, including dilutions such as palomino and spotting patterns such as pinto. Horses which are white in coat color are often mislabeled as "white" horses. However, a horse that looks white is usually a middle-aged or older gray. Grays are born a darker shade, get lighter as they age, and usually have black skin underneath their white hair coat (with the exception of pink skin under white markings). The only horses properly called white are born with a white hair coat and have predominantly pink skin, a fairly rare occurrence. There are no truly "albino" horses having both pink skin and red eyes.

Horse Breeding

Horse breeding is reproduction in horses, and particularly the human-directed process of selective breeding of animals, particularly purebred horses of a given breed. While feral and wild horses breed successfully without human assistance, planned matings can be used to produce specifically desired characteristics in domesticated horses. Furthermore, modern breeding management and technologies can increase the rate of conception, a healthy pregnancy, and successful foaling.

Terminology

The male parent of a horse, a stallion, is commonly known as the *sire* and the female parent, the mare, is called the *dam*. Both are genetically important, as each parent provides half of the genetic makeup of the ensuing offspring, called a foal. (Contrary to popular misuse, the word "colt" refers to a young male horse only; "filly" is a young female.) Though many horse owners may simply breed a family mare to a local stallion in order to produce a companion animal, most professional breeders use selective breeding to produce individuals of a given phenotype, or breed. Alternatively, a breeder could, using individuals of differing phenotypes, create a new breed with specific characteristics.

A horse is "bred" where it is foaled (born). Thus a foal conceived in England but foaled in the United States is regarded as being bred in the US. In some cases, most notably in the Thoroughbred breeding industry, American-bred horses may also be described by the state in which they are foaled. Some breeds denote the country, or state, where conception took place as the origin of the foal. Similarly, the "breeder", is the person who owned or leased the mare at the time of foaling. That individual may not have had anything to do with the mating of the mare. It's important to review each breed registry's rules to determine which applies to any specific foal.

In the horse breeding industry, the term "half-brother" or "half-sister" only describes horses which have the same dam, but different sires. Horses with the same sire but different dams are simply said to be "by the same sire", and no sibling relationship is implied. "Full" (or "own") siblings have both the same dam and the same sire. The terms paternal half-sibling, and maternal half-sibling are also often used. Three-quarter siblings are horses out of the same dam, and are by sires that are either half-brothers (i.e. same dam) or who are by the same sire.

Thoroughbreds and Arabians are also classified through the "distaff" or direct female line, known as their "family" or "tail female" line, tracing back to their taproot foundation bloodstock or the beginning of their respective stud books. The female line of descent always appears at the bottom of a tabulated pedigree and is therefore often known as the *bottom line*. "Linebreeding" technically is the duplication of fourth generation or more distant ancestors. However, the term is often used more loosely, describing horses with duplication of ancestors closer than the fourth generation. It also is sometimes used as a euphemism for the practice of inbreeding, a practice that is generally frowned upon by horse breeders, though used by some in an attempt to fix certain traits.

Estrous Cycle of the Mare

The estrous cycle (also spelled oestrous) controls when a mare is sexually receptive toward a stallion, and helps to physically prepare the mare for conception. It generally occurs during the spring and summer months, although some mares may be sexually receptive into the late fall, and is controlled by the photoperiod (length of the day), the cycle first triggered when the days begin to lengthen. The estrous cycle lasts about 19–22 days, with the average being 21 days. As the days shorten, the mare returns to a period when she is not sexually receptive, known as anestrus. Anestrus - occurring in the majority of, but not all, mares - prevents the mare from conceiving in the winter months, as that would result in her foaling during the harshest part of the year, a time when it would be most difficult for the foal to survive.

This cycle contains 2 phases:

- Estrus, or Follicular, phase: 5–7 days in length, when the mare is sexually receptive to a stallion. Estrogen is secreted by the follicle. Ovulation occurs in the final 24–48 hours of estrus.
- Diestrus, or Luteal, phase: 14–15 days in length, the mare is not sexually receptive to the stallion. The corpus luteum secretes progesterone.
- Depending on breed, on average, 16% of mares have double ovulations, allowing them to twin, this does not affect the length of time of estrus or diestrus.

Effects on the Reproductive System During the Estrous Cycle

Changes in hormone levels can have great effects on the physical characteristics of the reproductive organs of the mare, thereby preparing, or preventing, her from conceiving.

- Uterus: increased levels of estrogen during estrus cause edema within the uterus, making it feel heavier, and the uterus loses its tone. This edema decreases following ovulation, and the muscular tone increases. High levels of progesterone do not cause edema within the uterus. The uterus becomes flaccid during anestrus.
- Cervix: the cervix starts to relax right before estrus occurs, with maximal relaxation around the time of ovulation. The secretions of the cervix increase. High progesterone levels (during diestrus) cause the cervix to close and become toned.
- Vagina: the portion of the vagina near the cervix becomes engorged with blood right before estrus. The vagina becomes relaxed and secretions increase.
- Vulva: relaxes right before estrus begins. Becomes dry, and closes more tightly, during diestrus.

Hormones Involved in the Estrous Cycle, During Foaling and after Birth

The cycle is controlled by several hormones which regulate the estrous cycle, the mare's behaviour, and the reproductive system of the mare. The cycle begins when the increased day length causes the pineal gland to reduce the levels of melatonin, thereby allowing the hypothalamus to secrete GnRH.

- GnRH (Gonadotropin releasing hormone): secreted by the hypothalamus, causes the pituitary to release of 2 gonadotrophins: LH and FSH.
- LH (Luteinizing hormone): levels are highest 2 days following ovulation, then slowly decrease over 4–5 days, dipping to their lowest levels 5–16 days after ovulation. Stimulates the follicle to mature, which then in turn secretes estrogen. Unlike most mammals, the mare does not have an increase of LH right before ovulation.
- FSH (Follicle-stimulating hormone): secreted by the pituitary, causes the ovarian follicle to develop. Levels of FSH rise slightly at the end of estrus, but have their highest peak about 10 days before the next ovulation. FSH is inhibited by inhibin at the same time LH and estrogen levels rise, which prevents

immature follicles from continuing their growth. Mares may however have multiple FSH waves during a single estrous cycle, and diestrus follicles resulting from a diestrus FSH wave are not uncommon, particularly in the height of the natural breeding season.

- Estrogen: secreted by the developing follicle, it causes the pituitary gland to secrete more LH (therefore, these 2 hormones are in a positive feedback loop). Additionally, it causes behavioral changes in the mare, making her more receptive toward the stallion, and causes physical changes in the cervix, uterus, and vagina to prepare the mare for conception. Estrogen peaks 1–2 days before ovulation, and decreases within 2 days following ovulation.
- Inhibin: secreted by the developed follicle right before ovulation, "turns off" FSH, which is no longer needed now that the follicle is larger.
- Progesterone: prevents conception and decreases sexual receptibility of the mare to the stallion. Progesterone is therefore lowest during the estrus phase, and increases during diestrus. It decreases 12–15 days after ovulation, when the corpus luteum begins to decrease in size.
- Prostaglandin: secreted by the endrometrium 13–15 days following ovulation, causes luteolysis and prevents the corpus luteum from secreting progesterone
- eCG - equine chorionic gonadotropin - (also called PMSG (pregnant mare serum gonadotropin): chorionic gonadotropins secreted if the mare conceives. First secreted by the endometrial cups around the 36th day of gestation, peaking around day 60, and decreasing after about 120 days of gestation. Also help to stimulate the growth of the fetal gonads.
- Prolactin: stimulates lactation
- Oxytocin: stimulates the uterus to contract.

Breeding and Gestation

While horses in the wild mate and foal in mid to late spring, in the case of horses domestically bred for competitive purposes, especially horse racing and various futurities, it is desirable that they be born as close to January 1 in the northern hemisphere or August 1 in the southern hemisphere as possible, so as to be at an advantage in size and maturity when competing against other horses in the same age group. When an early foal is desired, barn managers will put the mare "under lights" by keeping the barn lights on in the winter to simulate a longer day, thus bringing the mare into estrus sooner than she would in nature. Mares signal estrus and ovulation by urination in the presence of a stallion, raising the tail and revealing the vulva. A stallion, approaching with a high head, will usually nicker, nip and nudge the mare, as well as sniff her urine to determine her readiness for mating.

Once fertilized, the oocyte (egg) remains in the oviduct for approximately 5.5 more days, and then descends into the uterus. The initial single cell combination is already dividing and by the time of entry into the uterus, the egg might have already reached the blastocyst stage.

The gestation period lasts for about eleven months, or about 340 days (normal average range 320–370 days). During the early days of pregnancy, the conceptus is mobile, moving about in the uterus until about day 16 when "fixation" occurs. Shortly after fixation, the embryo proper (so called up to about 35 days) will become visible on trans-rectal ultrasound (about day 21) and a heartbeat should be visible by about day 23. After the formation of the endometrial cups and early placentation is initiated (35–40 days of gestation) the terminology changes, and the embryo is referred to as a fetus. True implantation - invasion into the endometrium of any sort - does not occur until about day 35 of pregnancy with the formation of the endometrial cups, and true placentation (formation of the placenta) is not initiated until about day 40-45 and not completed until about 140 days of pregnancy. The fetus gender can be determined by day 70 of the gestation using ultrasound. Halfway through gestation the fetus is the size of between a rabbit and a beagle. The most dramatic fetal development occurs in the last 3 months of pregnancy when 60% of fetal growth occurs.

Care of the Pregnant Mare

Domestic mares receive specific care and nutrition to ensure that they and their foals are healthy. Mares are given vaccinations against diseases such as the Rhinopneumonitis (EHV-1) virus (which can cause abortions) as well as vaccines for other conditions that may occur in a given region of the world. Pre-foaling vaccines are recommended 4–6 weeks prior to foaling to maximize the immunoglobulin content of the colostrum in the first milk. Deworming the mare a few weeks prior to foaling is also important, as the mare is the primary source of parasites for the foal.

Mares can be used for riding or driving during most of their pregnancy, and it's healthy for them to have exercise. But only moderate exercise, especially when they become heavy in foal. Exercise in excessively high temperatures has been suggested as being detrimental to pregnancy maintenance during the embryonic period - it should however be noted that ambient temperatures encountered during the research were in the region of 100 degrees F and the same results may not be encountered in regions with lower ambient temperatures. During the last 3–4 months of gestation, rapid growth of the fetus increases the pregnant mare's nutritional requirements. Energy requirements during these last few months, and during the first few months of lactation are similar to those of a horse in full training. Trace minerals such as Copper are extremely important, particularly during the tenth month of pregnancy, for proper skeletal formation. Many feeds designed for pregnant and lactating mares provide the careful balance required of increased protein, increased calories through extra fat as well as vitamins and minerals. During the first several months of pregnancy, the nutritional requirements do not increase significantly since the rate of growth of the fetus is very slow. However, during this time, the mare should be provided supplemental vitamins and minerals, particularly if forage quality is questionable. Overfeeding the pregnant mare, particularly during early gestation, should be avoided, as excess weight may contribute to difficulties foaling or fetal/foal related problems.

Foaling

Mares due to foal are usually separated from other horses, both for the benefit of the mare and the safety of the soon-to-be-delivered foal. In addition, separation allows the mare to be monitored more closely by humans for any problems that may occur while giving birth. In the northern hemisphere a special foaling stall that is large and clutter free is frequently used, particularly by major breeding farms. Originally, this was due in part to a need for protection from the harsh winter climate present when mares foal early in the year, but even in moderate climates, such as Florida, foaling stalls are still common because they allow closer monitoring of mares. Smaller breeders often use a small pen with a large shed for foaling, or they may remove a wall between two box stalls in a small barn to make a large stall. In the milder climates seen in much of the southern hemisphere, most mares foal outside, often in a paddock built specifically for foaling, especially on the larger stud farms. Many stud farms worldwide employ technology to alert human managers when the mare is about to foal, including webcams, closed-circuit television, or assorted types of devices that alert a handler via a remote alarm when a mare lays down in a position to foal. On the other hand, some breeders, particularly those in remote areas or with extremely large numbers of horses, may allow mares to foal out in a field amongst a herd.

Most mares foal at night or early in the morning, and prefer to give birth alone when possible. Labour is rapid, often no more than 30 minutes, and from the time the feet of the foal appear to full delivery is often only about 15 to 20 minutes. Once the foal is born, the mare will lick the newborn foal to clean it and help blood circulation. In a very short time, the foal will attempt to stand and get milk from its mother. A foal should stand and nurse within the first hour of life.

To create a bond with her foal, the mare licks and nuzzles the foal, enabling her to distinguish hers from others. Some mares are aggressive when protecting their foals, and may attack other horses or unfamiliar humans that come near their newborns. After birth, a foal's navel is dipped in antiseptic to prevent infection, it is sometimes given an enema to help clear the meconium from its digestive tract, and the newborn is monitored to ensure that it stands and nurses without difficulty. While most horse births happen without complications, many owners have first aid supplies prepared and a veterinarian on call in case of a birthing emergency. People who supervise foaling should also watch the mare to be sure that she passes the placenta in a timely fashion, and that it is complete with no fragments remaining in the uterus, where retained fetal membranes could cause a serious inflammatory condition (endometritis) and/or infection. If the placenta is not removed from the stall after it is passed, a mare will often eat it, an instinct from the wild, where blood would attract predators.

Foal Care

Foals develop rapidly, and within a few hours a wild foal can travel with the herd. In domestic breeding, the foal and dam are usually separated from the herd for a while, but within a few weeks are typically pastured with the other horses. A foal will begin to eat hay, grass and grain alongside the mare at about 4 weeks old; by 10–12 weeks the foal requires more nutrition than the mare's milk can supply. Foals are typically weaned at 4–8 months of age, although in the wild a foal may nurse for a year.

How Breeds Develop

Beyond the appearance and conformation of a specific type of horse, breeders aspire to improve physical performance abilities. This concept, known as matching "form to function," has led to the development of not only different breeds, but also families or bloodlines within breeds that are specialists for excelling at specific tasks. For example, the Arabian horse of the desert naturally developed speed and endurance to travel long distances and survive in a harsh environment, and domestication by humans added a trainable disposition to the animal's natural abilities. In the meantime, in northern Europe, the locally adapted heavy horse with a thick, warm coat was domesticated and put to work as a farm animal that could pull a plow or wagon. This animal was later adapted through selective breeding to create a strong but ridable animal suitable for the heavily-armored knight in warfare.

Then, centuries later, when people in Europe wanted faster horses than could be produced from local horses through simple selective breeding, they imported Arabians and other oriental horses to breed as an outcross to the heavier, local animals. This led to the development of breeds such as the Thoroughbred, a horse taller than the Arabian and faster over the distances of a few miles required of a European race horse or light cavalry horse. Another cross between oriental and European horses produced the Andalusian, a horse developed in Spain that was powerfully built, but extremely nimble and capable of the quick bursts of speed over short distances necessary for certain types of combat as well as for tasks such as bullfighting.

Later, the people who settled the Americas needed a hardy horse that was capable of working with cattle. Thus, Arabians and Thoroughbreds were crossed on Spanish horses, both domesticated animals descended from those brought over by the Conquistadors, and feral horses such as the Mustangs, descended from the Spanish horse, but adapted by natural selection to the ecology and climate of the west. These crosses ultimately produced new breeds such as the American quarter horse and the Criollo of Argentina.

In modern times, these breeds themselves have since been selectively bred to further specialize at certain tasks. One example of this is the American quarter horse. Once a general-purpose working ranch horse, different bloodlines now specialize in different events. For example, larger, heavier animals with a very steady attitude are bred to give competitors an advantage in events such as team roping, where a horse has to start and stop quickly, but also must calmly hold a full-grown steer at the end of a rope. On the other hand, for an event known as cutting, where the horse must separate a cow from a herd and prevent it from rejoining the group, the best horses are smaller, quick, alert, athletic and highly trainable. They must learn quickly, have conformation that allows quick stops and fast, low turns, and the best competitors have a certain amount of independent mental ability to anticipate and counter the movement of a cow, popularly known as "cow sense."

Another example is the Thoroughbred. While most representatives of this breed are bred for horse racing, there are also specialized bloodlines suitable as show hunters or show jumpers. The hunter must have a tall, smooth build that allows it to trot and canter smoothly and efficiently. Instead of speed, value is placed on appearance and upon giving the equestrian a comfortable ride, with natural jumping ability that shows bascule and good form. A show jumper, however, is bred less for overall form and more for power over tall fences, along with speed, scope, and agility. This favours a horse with a good galloping stride, powerful hindquarters that can change speed or direction easily, plus a good shoulder angle and length of neck. A jumper has a more powerful build than either the hunter or the racehorse.

History of Horse Breeding

The history of horse breeding goes back millennia. Though the precise date is in dispute, humans could have domesticated the horse as far back as approximately 4500 BCE. However, evidence of planned breeding has a more blurry history. One of the earliest people known to document the breedings of their horses were the Bedouin of the Middle East, the breeders of the Arabian horse. While it is difficult to determine how far back the Bedouin passed on pedigree information via an oral tradition, there were written pedigrees of Arabian horses by A.D. 1330. The Akhal-Teke of West-Central Asia is another breed with roots in ancient times that was also bred specifically for war and racing. The nomads of the Mongolian steppes bred horses for several thousand years as well. The types of horses bred varied with culture and with the times. The uses to which a horse was put also determined its qualities, including smooth amblers for riding, fast horses for carrying messengers, heavy horses for plowing and pulling heavy wagons, ponies for hauling cars of ore from mines, packhorses, carriage horses and many others.

Medieval Europe bred large horses specifically for war, called destriers. These horses were the ancestors of the great heavy horses of today, and their size was preferred not simply because of the weight of the armor, but also because a large horse provided more power for the knight's lance. Weighing almost twice as much as a normal riding horse, the destrier was a powerful weapon in battle.

On the other hand, during this same time, lighter horses were bred in northern Africa and the Middle East by Muslim warriors, who preferred a faster, more agile horse. The lighter horse suited the raids and battles of the Bedouins, allowing them to outmaneuver rather than overpower the enemy. When Muslim warriors and European knights collided in warfare, the heavy knights were frequently outmaneuvered. The Europeans, however, soon made up for the lack of speed of their native breeds by incorporating genetic traits from captured oriental horses such as the Arabian, Barb to their stables. This cross-breeding led both to a nimbler war horse, such as today's Percheron, but also to created a type of horse known as a Courser, a predecessor to the Thoroughbred, which was used as a message horse.

During the Renaissance, horses were bred not only for war, but for haute ecole riding, derived from the most athletic movements required of a war horse, and popular among the elite nobility of the time. Breeds such as the Lipizzan were developed from Spanish-bred horses for this purpose, and also became the preferred mounts of cavalry officers, who were derived mostly from the ranks of the nobility. It was during this time that gunpowder was developed, and so the light cavalry horse, a faster and quicker war horse, was bred for a "shoot and run" tactic rather than the close hand-to-hand fighting seen in the Middle Ages.

After Charles II retook the British throne in 1660, horse racing, which had been banned by Cromwell, was revived. The Thoroughbred was developed 40 years later, bred to be the ultimate racehorse, through the lines of 3 foundation Arabian stallions. In the 18th century, James Burnett, Lord Monboddo noted the importance of selecting appropriate parentage to achieve desired outcomes of successive generations. Monboddo worked more broadly in the abstract thought of species relationships and evolution of species. The Thoroughbred breeding hub in Lexington, Kentucky was developed in the late 18th century, and became a mainstay in American racehorse breeding.

The 17th and 18th centuries saw more of a need for fine carriage horses in Europe, bringing in the dawn of the warmblood. The warmblood breeds have been exceptionally good at adapting to changing times, and from their carriage horse beginnings they easily transitioned during the 20th century into a sport horse type. Today's warmblood breeds, although still used for competitive driving, are more often seen competing in the show jumping or dressage arenas. The Thoroughbred continues to dominate the horseracing world, although its lines have been more recently used to improve warmblood breeds and to develop sport horses.

The predecessor of the American Quarter Horse was developed in the 18th century, mainly for quarter racing (racing ¼ of a mile). The breed was later adapted for work in the west, and "cow sense" was particularly bred for as their use for herding cattle increased. However, because there was also a need for animals suitable for sprint racing, the modern Quarter Horse has two distinct types: the sleeker racing type and the stock horse type. The racing type most resembles the finer-boned ancestors of the first racing Quarter Horses, and the type is still used for ¼-mile races. The stock horse type, used in western events, is bred for a shorter stride, docile temperament, and cow sense.

The need for horses for heavy draft and carriage work continued until the industrial revolution and the advent of the automobile and the tractor. After this time, draft and carriage horse numbers dropped significantly, though light riding horses remained popular for recreational pursuits. Draft horses today are used on a few small farms, but today are seen mainly for pulling and plowing competitions rather than farm work. Heavy harness horses are now used as an outcross with lighter breeds, such as the Thoroughbred, to produce the modern warmblood breeds popular in Olympic and sport horse disciplines.

Deciding to Breed a Horse

Breeding a horse is an endeavor where the owner, particularly of the mare, will usually need to invest considerable time and money. For this reason, a horse owner needs to consider several factors, including:

- Does the proposed breeding animal have valuable genetic qualities to pass on?
- Is the proposed breeding animal in good physical health, fertile, and able to withstand the vigors of reproduction?
- For what purpose will the foal be used?
- Is there a market for the foal in the event that the owner does not wish to keep the foal for its entire life?
- What is the anticipated economic benefit, if any, to the owner of the ensuing foal?
- What is the anticipated economic benefit, if any, to the owner(s) of the sire and dam or the foal?
- Does the owner of the mare have the expertise to properly manage the mare through gestation and parturition?
- Does the owner of the potential foal have the expertise to properly manage and train a young animal once it is born?

There are value judgements involved in considering whether an animal is suitable breeding stock, hotly debated by breeders. Additional personal beliefs may come into play when considering a suitable level of care for the mare and ensuing foal, the potential market or use for the foal, and other tangible and intangible benefits to the owner.

If the breeding endeavor is intended to make a profit, there are additional market factors to consider, which may vary considerably from year to year, from breed to breed, and by region of the world. In many cases, the low end of the market is saturated with horses, and the law of supply and demand thus allows little or no profit to be made from breeding unregistered animals or animals of poor quality, even if registered. The minimum cost of breeding for a mare owner includes the stud fee, and the cost of proper nutrition, management and veterinary care of the mare throughout gestation, parturition, and care of both mare and foal up to the time of weaning. Veterinary expenses may be higher if specialized reproductive technologies are used or health complications occur.

Making a profit in horse breeding is often difficult. While some owners of only a few horses may keep a foal for purely personal enjoyment, many individuals breed horses in hopes of making some money in the process. A general rule of thumb is that a foal intended for sale should be worth three times the cost of the stud fee if it were sold at the moment of birth. From birth forward, the costs of care and training are added to the value of the foal, with a sale price going up accordingly. If the foal wins awards in some form of competition, that may also enhance the price.

On the other hand, without careful thought, foals bred without a potential market for them may wind up being sold at a loss, and in a worst-case scenario, sold for "salvage" value—a euphemism for sale to slaughter as horsemeat.

Therefore, a mare owner must consider their reasons for breeding, asking hard questions of themselves as to whether their motivations are based on either emotion or profit and how realistic those motivations may be.

Choosing Breeding Stock

The stallion should be chosen to complement the mare, with the goal of producing a foal that has the best qualities of both animals, yet avoids having the weaker qualities of either parent. Generally, the stallion should have proven himself in the discipline or sport the mare owner wishes for the "career" of the ensuing foal. Mares should also have a competition record showing that they also have suitable traits, though this does not happen as often.

Some breeders consider the quality of the sire to be more important than the quality of the dam. However, other breeders maintain that the mare is the most important parent. Because stallions can produce far more offspring than mares, a single stallion can have a greater overall impact on a breed. However, the mare may have a greater influence on an individual foal because its physical characteristics influence the developing foal in the womb and the foal also learns habits from its dam when young. Foals may also learn the "language of intimidation and submission" from their dam, and this imprinting may affect the foal's status and rank within the herd. Many times, a mature horse will achieve status in a herd similar to that of its dam; the offspring of dominant mares become dominant themselves.

A purebred horse is usually worth more than a horse of mixed breeding, though this matters more in some disciplines than others. The breed of the horse is sometimes secondary when breeding for a sport horse, but some disciplines may prefer a certain breed or a specific phenotype of horse. Sometimes, purebred bloodlines are an absolute requirement: For example most Racehorses in the world must be recorded with a breed registry in order to race.

Bloodlines are often considered, as some bloodlines are known to cross well with others. If the parents have not yet proven themselves by competition or by producing quality offspring, the bloodlines of the horse are often a good indicator of quality and possible strengths and weaknesses. Some bloodlines are known not only for their athletic ability, but could also carry a conformational or genetic defect, poor temperament, or for a medical problem. Some bloodlines are also fashionable or otherwise marketable, which is an important consideration should the mare owner wish to sell the foal.

Horse breeders also consider conformation, size and temperament. All of these traits are heritable, and will determine if the foal will be a success in its chosen discipline. The offspring, or "get", of a stallion are often excellent indicators of his ability to pass on his characteristics, and the particular traits he actually passes on. Some stallions are fantastic performers but never produce offspring of comparable quality. Others sire fillies of great abilities but not colts. At times, a horse of mediocre ability sires foals of outstanding quality.

Mare owners also look into the question of if the stallion is fertile and has successfully "settled" (i.e. impregnated) mares. A stallion may not be able to breed naturally, or old age may decrease his performance. Mare care boarding fees and semen collection fees can be a major cost.

Costs Related to Breeding

Breeding a horse can be an expensive endeavor, whether breeding a backyard competition horse or the next Olympic medalist. Costs may include:

- The stud and booking fee
- Fees for collecting, handling, and transporting semen (if AI is used and semen is shipped)
- Mare exams: to determine if she is healthy enough to breed, to determine when she ovulates, and (if AI is used) to inseminate her
- Mare transport, care, and board if the mare is bred live cover at the stallion's residence
- Veterinary bills to keep the pregnant mare healthy while in foal
- Possible veterinary bills during pregnancy or foaling should something go wrong
- Veterinary bills for the foal for its first exam a few days following foaling.

Stud fees are determined by the quality of the stallion, his performance record, the performance record of his get (offspring), as well as the sport and general market that the animal is standing for.

The highest stud fees are generally for racing Thoroughbreds, which may charge from two to three thousand dollars for a breeding to a new or unproven stallion, to several hundred thousand dollars for a breeding to a proven producer of stakes winners. Stallions in other disciplines often have stud fees that begin in the range of $1000 to $3000, with top contenders who produce champions in certain disciplines able to command as much as $20,000 for one breeding. The lowest stud fees to breed to a grade horse or an animal of low-quality pedigree may only be $100–$200, but there are trade-offs: the horse will probably be unproven, and likely to produce lower-quality offspring than a horse with a stud fee that is in the typical range for quality breeding stock.

As a stallion's career, either performance or breeding, improves, his stud fee tends to increase in proportion. If one or two offspring are especially successful, winning several stakes races or an Olympic medal, the stud fee will generally greatly increase. Younger, unproven stallions will generally have a lower stud fee earlier on in their careers.

To help decrease the risk of financial loss should the mare die or abort the foal while pregnant, many studs have a live foal guarantee (LFG) - also known as "no foal, free return" or "NFFR" - allowing the owner to have a free breeding to their stallion the next year. However, this is not offered for every breeding.

Covering the Mare

There are two general ways to "cover" or breed the mare:

- Live cover: the mare is brought to the stallion's residence and is covered "live" in the breeding shed. She may also be turned out in a pasture with the stallion for several days to breed naturally ('pasture bred'). The former situation is often preferred, as it provides a more controlled environment, allowing the breeder to ensure that the mare was covered, and places the handlers in a position to remove the horses from one another should one attempt to kick or bite the other.
- Artificial Insemination (AI): the mare is inseminated by a veterinarian or an equine reproduction manager, using either fresh, cooled or frozen semen.

After the mare is bred or artificially inseminated, she should be checked using ultrasound 14–16 days later. If she "took", and is pregnant. A second check is usually performed at 28 days. If the mare is not pregnant, she may be bred again during her next cycle.

It is considered safe to breed a mare to a stallion of much larger size. Because of the mare's type of placenta and its attachment and blood supply, the foal will be limited in its growth within the uterus to the size of the mare's uterus, but will grow to its genetic potential after it is born. Test breedings have been done with draft horse stallions bred to small mares with no increase in the number of difficult births.

Live Cover

When breeding live cover, the mare is usually boarded at the stud. She is "teased" several times with a stallion that will not breed to her, usually with the stallion being presented to the mare over a barrier. Her reaction to the teaser, whether hostile or passive, is noted.

A mare that is in heat will generally tolerate a teaser (although this is not always the case), and may present herself to him, holding her tail to the side. A veterinarian may also determine if the mare is ready to be bred, by ultrasound or palpating daily to determine if ovulation has occurred.

When it has been determined that the mare is ready, both the mare and intended stud will be cleaned. The mare will then be presented to the stallion, usually with one handler controlling the mare and one or more handlers in charge of the stallion. Multiple handlers are preferred, as the mare and stallion can be easily separated should there be any trouble. The Jockey Club, the organization that oversees the Thoroughbred industry in the United States, requires all registered foals to be bred through live cover. Artificial insemination, listed below, is not permitted. Similar rules apply in other countries.

By contrast, the U.S. standardbred industry allows registered foals to be bred by live cover, or by artificial insemination (AI) with fresh or frozen (not dried) semen. No other artificial fertility treatment is allowed. In addition, foals bred via AI of frozen semen may only be registered if the stallion's sperm was collected during his lifetime, and used no later than the calendar year of his death or castration.

Artificial Insemination

Artificial insemination (AI) has several advantages over live cover, and has a very similar conception rate:

- The mare and stallion never have to come in contact with each other, which therefore reduces breeding accidents, such as the mare kicking the stallion.
- AI opens up the world to international breeding, as semen may be shipped across continents to mares that would otherwise be unable to breed to a particular stallion.
- A mare also does not have to travel to the stallion, so the process is less stressful on her, and if she already has a foal, the foal does not have to travel.
- AI allows more mares to be bred from one stallion, as the ejaculate may be split between mares.
- AI reduces the chance of spreading sexually transmitted diseases between mare and stallion.
- AI allows mares or stallions with health issues, such as sore hocks which may prevent a stallion from mounting, to continue to breed.
- Frozen semen may be stored and used to breed mares even after the stallion is dead, allowing his lines to continue. However, the semen of some stallions does not freeze well. Some breed registries may not permit the registration of foals resulting from the use of frozen semen after the stallion's death,

although other large registries accept such usage and provide registrations. The overall trend is toward permitting use of frozen semen after the death of the stallion.

A stallion is usually trained to mount a phantom (or dummy) mare, although a live mare may be used, and he is most commonly collected using an artificial vagina (AV) which is heated to simulate the vagina of the mare. The AV has a filter and collection area at one end to capture the semen, which can then be processed in a lab. The semen may be chilled or frozen and shipped to the mare owner or used to breed mares "on-farm". When the mare is in heat, the person inseminating introduces the semen directly into her uterus using a syringe and pipette.

Advanced Reproductive Techniques

Often an owner does not want to take a valuable competition mare out of training to carry a foal. This presents a problem, as the mare will usually be quite old by the time she is retired from her competitive career, at which time it is more difficult to impregnate her. Other times, a mare may have physical problems that prevent or discourage breeding. However, there are now several options for breeding these mares. These options also allow a mare to produce multiple foals each breeding season, instead of the usual one. Therefore, mares may have an even greater value for breeding.

- Embryo Transfer: This relatively new method involves flushing out the mare's fertilized embryo a few days following insemination, and transferring to a surrogate mare, which has been synchronized to be in the same phase of the estrous cycle as the donor mare.
- Gamete Intrafallopian Transfer (GIFT): The mare's ovum and the stallion's sperm are deposited in the oviduct of a surrogate dam. This technique is very useful for subfertile stallions, as fewer sperm are needed, so a stallion with a low sperm count can still successfully breed.
- Egg Transfer: An oocyte is removed from the mare's follicle and transferred into the oviduct of the recipient mare, who is then bred. This is best for mares with physical problems, such as an obstructed oviduct, that prevent breeding.

Reproduction and Development

Gestation lasts for approximately 335–340 days and usually results in one foal. Twins are rare. Horses are a precocial species, and foals are capable of standing and running within a short time following birth.

Horses, particularly colts, sometimes are physically capable of reproduction at about 18 months, but domesticated horses are rarely allowed to breed before the age of three, especially females.

Horses four years old are considered mature, although the skeleton normally continues to develop until the age of six; maturation also depends on the horse's size, breed, sex, and quality of care. Also, if the horse is larger, its bones are larger; therefore, not only do the bones take longer to actually form bone tissue, but the epiphyseal plates are also larger and take longer to convert from cartilage to bone. These plates convert after the other parts of the bones, and are crucial to development. Depending on maturity, breed, and work expected, horses are usually put under saddle and trained to be ridden between the ages of two and four. Although Thoroughbred race horses are put on the track at as young as two years old in some countries, horses specifically bred for sports such as dressage are generally not put under saddle until they are three or four years old, because their bones and muscles are not solidly developed.

For endurance riding competition, horses are not deemed mature enough to compete until they are a full 60 calendar months (5 years) old.

Interaction with Humans

Worldwide, horses play a role within human cultures and have done so for millennia. Horses are used for leisure activities, sports, and working purposes. The Food and Agriculture Organization (FAO) estimates that in 2008, there were almost 59,000,000 horses in the world, with around 33,500,000 in the Americas, 13,800,000 in Asia and 6,300,000 in Europe and smaller portions in Africa and Oceania.

There are estimated to be 9,500,000 horses in the United States alone. The American Horse Council estimates that horse-related activities have a direct impact on the economy of the United States of over $39 billion, and when indirect spending is considered, the impact is over $102 billion. In a 2004 "poll" conducted by Animal Planet, more than 50,000 viewers from 73 countries voted for the horse as the world's 4th favorite animal.

Communication between human and horse is paramount in any equestrian activity; to aid this process horses are usually ridden with a saddle on their backs to assist the rider with balance and positioning, and a bridle or related headgear to assist the rider in maintaining control. Sometimes horses are ridden without a saddle, and occasionally, horses are trained to perform without a bridle or other headgear. Many horses are also driven, which requires a harness, bridle, and some type of vehicle.

Sport

Historically, equestrians honed their skills through games and races. Equestrian sports provided entertainment for crowds and honed the excellent horsemanship that was needed in battle. Many sports, such as dressage, eventing and show jumping, have origins in military training, which were focused on control and balance of both horse and rider.

Other sports, such as rodeo, developed from practical skills such as those needed on working ranches and stations. Sport hunting from horseback evolved from earlier practical hunting techniques. Horse racing of all types evolved from impromptu competitions between riders or drivers. All forms of competition, requiring demanding and specialized skills from both horse and rider, resulted in the systematic development of specialized breeds and equipment for each sport. The popularity of equestrian sports through the centuries has resulted in the preservation of skills that would otherwise have disappeared after horses stopped being used in combat.

Horses are trained to be ridden or driven in a variety of sporting competitions. Examples include show jumping, dressage, three-day eventing, competitive driving, endurance riding, gymkhana, rodeos, and fox hunting. Horse shows, which have their origins in medieval European fairs, are held around the world. They host a huge range of classes, covering all of the mounted and harness disciplines, as well as "In-hand" classes where the horses are led, rather than ridden, to be evaluated on their conformation. The method of judging varies with the discipline, but winning usually depends on style and ability of both horse and rider. Sports such as polo do not judge the horse itself, but rather use the horse as a partner for human competitors as a necessary part of the game.

Although the horse requires specialized training to participate, the details of its performance are not judged, only the result of the rider's actions—be it getting a ball through a goal or some other task. Examples of these sports of partnership between human and horse include jousting, in which the main goal is for one rider to unseat the other, and buzkashi, a team game played throughout Central Asia, the aim being to capture a goat carcass while on horseback.

Horse racing is an equestrian sport and major international industry, watched in almost every nation of the world. There are three types: "flat" racing; steeplechasing, i.e. racing over jumps; and harness racing, where horses trot or pace while pulling a driver in a small, light cart known as a sulky. A major part of horse racing's economic importance lies in the gambling associated with it.

Work

There are certain jobs that horses do very well, and no technology has yet developed to fully replace them. For example, mounted police horses are still effective for certain types of patrol duties and crowd control. Cattle ranches still require riders on horseback to round up cattle that are scattered across remote, rugged terrain. Search and rescue organizations in some countries depend upon mounted teams to locate people, particularly hikers and children, and to provide disaster relief assistance. Horses can also be used in areas where it is necessary to avoid vehicular disruption to delicate soil, such as nature reserves. They may also be the only form of transport allowed in wilderness areas. Horses are quieter than motorized vehicles. Law enforcement officers such as park rangers or game wardens may use horses for patrols, and horses or mules may also be used for clearing trails or other work in areas of rough terrain where vehicles are less effective.

Although machinery has replaced horses in many parts of the world, an estimated 100 million horses, donkeys and mules are still used for agriculture and transportation in less developed areas. This number includes around 27 million working in Africa alone. Some land management practices such as cultivating and logging can be efficiently performed with horses. In agriculture, less fossil fuel is used and increased environmental conservation occurs over time with the use of draft animals such as horses. Logging with horses can result in reduced damage to soil structure and less damage to trees due to more selective logging.

Entertainment and Culture

Modern horses are often used to reenact many of their historical work purposes. Horses are used, complete with equipment that is authentic or a meticulously recreated replica, in various live action historical reenactments of specific periods of history, especially recreations of famous battles. Horses are also used to preserve cultural traditions and for ceremonial purposes.

Countries such as the United Kingdom still use horse-drawn carriages to convey royalty and other VIPs to and from certain culturally significant events. Public exhibitions are another example, such as the Budweiser Clydesdales, seen in parades and other public settings, a team of draft horses that pull a beer wagon similar to that used before the invention of the modern motorized truck.

Horses are frequently seen in television and films. They are used both as main characters, in films such as *Seabiscuit*, and *Dreamer*, and as visual elements that assure the accuracy of historical stories. Both live horses and iconic images of horses are used in advertising to promote a variety of products. The horse frequently appears in coats of arms in heraldry.

The horse can be represented as standing, walking (passant), trotting, running (courant), rearing (rampant or forcine) or springing (salient). The horse may be saddled and bridled, harnessed, or without any apparel whatsoever. The horse also appears in the 12-year cycle of animals in the Chinese zodiac related to the Chinese calendar. According to Chinese folklore, each animal is associated with certain personality traits, and those born in the year of the horse are intelligent, independent, and free-spirited.

Therapeutic Use

People of all ages with physical and mental disabilities obtain beneficial results from association with horses. Therapeutic riding is used to mentally and physically stimulate disabled persons and help them improve their lives through improved balance and coordination, increased self-confidence, and a greater feeling of freedom and independence. The benefits of equestrian activity for people with disabilities has also been recognized with the addition of equestrian events to the Paralympic Games and recognition of para-equestrian events by the International Federation for Equestrian Sports (FEI). Hippotherapy and therapeutic horseback riding are names for different physical, occupational, and speech therapy treatment strategies that utilize equine movement. In hippotherapy, a therapist uses the horse's movement to improve their patient's cognitive, coordination, balance, and fine motor skills, whereas therapeutic horseback riding uses specific riding skills.

Horses also provide psychological benefits to people whether they actually ride or not. "Equine-assisted" or "equine-facilitated" therapy is a form of experiential psychotherapy that uses horses as companion animals to assist people with mental illness, including anxiety disorders, psychotic disorders, mood disorders, behavioral difficulties, and those who are going through major life changes. There are also experimental programs using horses in prison settings. Exposure to horses appears to improve the behaviour of inmates and help reduce recidivism when they leave.

Warfare

Horses in warfare have been seen for most of recorded history. The first archaeological evidence of horses used in warfare dates to between 4000 to 3000 BC, and the use of horses in warfare was widespread by the end of the Bronze Age. Although mechanization has largely replaced the horse as a weapon of war, horses are still seen today in limited military uses, mostly for ceremonial purposes, or for reconnaissance and transport activities in areas of rough terrain where motorized vehicles are ineffective. Horses have been used in the 21st century by the Janjaweed militias in the War in Darfur.

Products

Horses are raw material for many products made by humans throughout history, including byproducts from the slaughter of horses as well as materials collected from living horses. Products collected from living horses include mare's milk, used by people with large horse herds, such as the Mongols, who let it ferment to produce kumis. Horse blood was once used as food by the Mongols and other nomadic tribes, who found it a convenient source of nutrition when traveling.

Drinking their own horses' blood allowed the Mongols to ride for extended periods of time without stopping to eat. Today, the drug Premarin is a mixture of estrogens extracted from the urine of pregnant mares (pregnant mares' urine). It is a widely used drug for hormone replacement therapy. The tail hair of horses can be used for making bows for string instruments such as the violin, viola, cello, and double bass.

Horse meat has been used as food for humans and carnivorous animals throughout the ages. It is eaten in many parts of the world, though consumption is taboo in some cultures. Horsemeat has been an export industry in the United States and other countries, though legislation has periodically been introduced in the United States Congress which would end export from the United States. Horsehide leather has been used for boots, gloves, jackets, baseballs, and baseball gloves. Horse hooves can also be used to produce animal glue. Horse bones can be used to make implements. Specifically, in Italian cuisine, the horse tibia is sharpened into a probe called a *spinto*, which is used to test the readiness of a (pig) ham as it cures. In Asia, the saba is a horsehide vessel used in the production of kumis.

Care

Horses are grazing animals, and their major source of nutrients is good-quality forage from hay or pasture. They can consume approximately 2% to 2.5% of their body weight in dry feed each day. Therefore, a 450-kilogram (990 lb) adult horse could eat up to 11 kilograms (24 lb) of food. Sometimes, concentrated feed such as grain is fed in addition to pasture or hay, especially when the animal is very active. When grain is fed, equine nutritionists recommend that 50% or more of the animal's diet by weight should still be forage.

Horses require a plentiful supply of clean water, a minimum of 10 US gallons (38 L) to 12 US gallons (45 L) per day. Although horses are adapted to live outside, they require shelter from the wind and precipitation, which can range from a simple shed or shelter to an elaborate stable.

Horses require routine hoof care from a farrier, as well as vaccinations to protect against various diseases, and dental examinations from a veterinarian or a specialized equine dentist. If horses are kept inside in a barn, they require regular daily exercise for their physical health and mental well-being. When turned outside, they require well-maintained, sturdy fences to be safely contained. Regular grooming is also helpful to help the horse maintain good health of the hair coat and underlying skin.

Chapter 9

Sheep and Goat Production

Sheep (*Ovis aries*) are quadrupedal, ruminant mammals typically kept as livestock. Like all ruminants, sheep are members of the order Artiodactyla, the even-toed ungulates. Although the name "sheep" applies to many species in the genus *Ovis*, in everyday usage it almost always refers to *Ovis aries*. Numbering a little over one billion, domestic sheep are also the most numerous species of sheep.

Sheep are most likely descended from the wild mouflon of Europe and Asia. One of the earliest animals to be domesticated for agricultural purposes, sheep are raised for fleece, meat (lamb, hogget or mutton) and milk. A sheep's wool is the most widely used animal fibre, and is usually harvested by shearing. Ovine meat is called lamb when from younger animals and mutton when from older ones. Sheep continue to be important for wool and meat today, and are also occasionally raised for pelts, as dairy animals, or as model organisms for science.

Sheep husbandry is practised throughout the majority of the inhabited world, and has been fundamental to many civilizations. In the modern era, Australia, New Zealand, the southern and central South American nations, and the British Isles are most closely associated with sheep production.

Sheep-raising has a large lexicon of unique terms which vary considerably by region and dialect. Use of the word *sheep* began in Middle English as a derivation of the Old English word *scçap*; it is both the singular and plural name for the animal. A group of sheep is called a flock, herd or mob. Adult female sheep are referred to as ewes, intact males as rams or occasionally tups, castrated males as wethers, and younger sheep as lambs. Many other specific terms for the various life stages of sheep exist, generally related to lambing, shearing, and age. Being a key animal in the history of farming, sheep have a deeply entrenched place in human culture, and find representation in much modern language and symbology. As livestock, sheep are most-often associated with pastoral, Arcadian imagery.

Description

Domestic sheep are relatively small ruminants, usually with a crimped hair called wool and often with horns forming a lateral spiral. Domestic sheep differ from their wild relatives and ancestors in several respects, having become uniquely neotenic as a result of selective breeding by humans. A few primitive breeds of sheep retain some of the characteristics of their wild cousins, such as short tails. Depending on breed, domestic sheep may have no horns at all (polled), or horns in both sexes (as in wild sheep), or in males only. Most horned breeds have a single pair, but a few breeds may have several.

Another trait unique to domestic sheep (as compared to wild ovines, not other livestock) is their wide variation in color. Wild sheep are largely variations of brown hues, and variation with species is extremely limited. Colours of domestic sheep range from pure white to dark chocolate brown and even spotted or piebald. Selection for easily dyeable white fleeces began early in sheep domestication, and as white wool is a dominant trait it spread quickly. However, colored sheep do appear in many modern breeds, and may even appear as a recessive trait in white flocks. While white wool is desirable for large commercial markets, there is a niche market for colored fleeces, mostly for handspinning. The nature of the fleece varies widely among the breeds, from dense and highly crimped, to long and hair-like. There is variation of wool type and quality even among members of the same flock, so wool classing is a step in the commercial processing of the fibre.

Suffolks are a medium wool, black-faced breed of meat sheep that make up 60% of the sheep population in the U.S.

Depending on breed, sheep show a range of heights and weights. Their rate of growth and mature weight is a heritable trait that is often selected for in breeding. Ewes typically weigh between 45 and 100 kilograms (99 and 220 lb), and rams between 45 and 160 kilograms (99 and 350 lb). Mature sheep have 32 teeth. As with other ruminants, the front teeth in the lower jaw bite against a hard, toothless pad in the upper jaw. These are used to pick off vegetation, then the rear teeth grind it before it is swallowed. There are eight lower front teeth in ruminants, but there is some disagreement as to whether these are eight incisors, or six incisors and two incisor-shaped canines. This means that the dental formula for sheep is either I:0/4 C:0/0 P:3/3 M:3/3, or I:0/3 C:0/1 P:3/3 M:3/3. There is a large toothless gap between the front "biting" teeth and the rear "grinding" teeth.

For the first few years of life it is possible to calculate the age of sheep from their front teeth, as a pair of milk teeth is replaced by larger adult teeth each year, the full set of eight adult front teeth being complete at about four years of age. The front teeth are then gradually lost as sheep age, making it harder for them to feed and hindering the health and productivity of the animal. For this reason, domestic sheep on normal pasture begin to slowly decline from four years on, and the average life expectancy of a sheep is 10 to 12 years, though some sheep may live as long as 20 years.

Sheep have good hearing, and are sensitive to noise when being handled. Sheep have horizontal slit-shaped pupils, possessing excellent peripheral vision; with visual fields of approximately 270° to 320°, sheep can behind themselves without turning their heads. However, sheep have poor depth perception; shadows and dips in the ground may cause sheep to baulk. In general, sheep have a tendency to move out of the dark and into well-lit areas, and prefer to move uphill when disturbed. Sheep also have an excellent sense of smell, and, like all species of their genus, have scent glands just in front of the eyes, and interdigitally on the feet. The purpose of these glands is uncertain, but those on the face may be used in breeding behaviours. The foot glands might also be related to reproduction, but alternative reasons, such as secretion of a waste product or a scent marker to help lost sheep find their flock, have also been proposed.

Sheep and goats are closely related as both are in the subfamily Caprinae. However, they are separate species, so hybrids rarely occur, and are always infertile. A hybrid of a ewe and a buck (a male goat) is called a sheep-goat hybrid (only a single such animal has been confirmed), and is not to be confused with the genetic chimera called a geep. Visual differences between sheep and goats include the beard and divided upper lip unique to goats. Sheep tails also hang down, even when short or docked, while the short tails of goats are held upwards. Sheep breeds are also often naturally polled (either in both sexes or just in the female), while naturally polled goats are rare (though many are polled artificially). Males of the two species differ in that buck goats acquire a unique and strong odour during the rut, whereas rams do not.

Breeds

The domestic sheep is a multi-purpose animal, and the more than 200 breeds now in existence were created to serve these diverse purposes. Some sources give a count of a thousand or more breeds, but these numbers cannot be verified. Almost all sheep are classified as being best suited to furnishing a certain product: wool, meat, milk, hides, or a combination in a dual-purpose breed. Other features used when classifying sheep include face color (generally white or black), tail length, presence or lack of horns, and the topography for which the breed has been developed. This last point is especially stressed in the UK, where breeds are described as either upland (hill or mountain) or lowland breeds. A sheep may also be of a fat-tailed type, which is a dual-purpose sheep common in Africa and Asia with larger deposits of fat within and around its tail.

Breeds are also grouped based on how well they are suited to producing a certain type of breeding stock. Generally, sheep are thought to be either "ewe breeds" or "ram breeds". Ewe breeds are those that are hardy, and have good reproductive and mothering capabilities – they are for replacing breeding ewes in standing flocks. Ram breeds are selected for rapid growth and carcase quality, and are mated with ewe breeds to produce meat lambs. Lowland and upland breeds are also crossed in this fashion, with the hardy hill ewes crossed with larger, fast-growing lowland rams to produce ewes called mules, which can then be crossed with meat-type rams to produce prime market lambs. Many breeds, especially rare or primitive ones, fall into no clear category.

The Barbados Blackbelly is a Hair Sheep Breed of Caribbean Origin

Breeds are categorized by the type of their wool. Fine wool breeds are those that have wool of great crimp and density, which are preferred for textiles. Most of these were derived from Merino sheep, and the breed continues to dominate the world sheep industry. Downs breeds have wool between the extremes, and are typically fast-growing meat and ram breeds with dark faces. Some major medium wool breeds, such as the Corriedale, are dual-purpose crosses of long and fine-wooled breeds and were created for high-production commercial flocks. Long wool breeds are the largest of sheep, with long wool and a slow rate of growth. Long wool sheep are most valued for crossbreeding to improve the attributes of other sheep types. For example: the American Columbia breed was developed by crossing Lincoln rams (a long wool breed) with fine-wooled Rambouillet ewes.

Coarse or carpet wool sheep are those with a medium to long length wool of characteristic coarseness. Breeds traditionally used for carpet wool show great variability, but the chief requirement is a wool that will not break down under heavy use (as would that of the finer breeds). As the demand for carpet-quality wool declines, some breeders of this type of sheep are attempting to use a few of these traditional breeds for alternative purposes. Others have always been primarily meat-class sheep. A minor class of sheep are the dairy breeds. Dual-purpose breeds that may primarily be meat or wool sheep are often used secondarily as milking animals, but there are a few breeds that are predominantly used for milking. These sheep do produce a higher quantity of milk and have slightly longer lactation curves. In the quality of their milk, fat and protein content percentages of dairy sheep vary from non-dairy breeds but lactose content does not.

A last group of sheep breeds is that of fur or hair sheep, which do not grow wool at all. Hair sheep are similar to the early domesticated sheep kept before woolly breeds were developed, and are raised for meat and pelts. Some modern breeds of hair sheep, such as the Dorper, result from crosses between wool and hair breeds. For meat and hide producers, hair sheep are cheaper to keep, as they do not need shearing. Hair sheep are also more resistant to parasites and hot weather.

With the modern rise of corporate agribusiness and the decline of localized family farms, many breeds of sheep are in danger of extinction. The Rare Breeds Survival Trust of the UK lists 22 native breeds as having only 3,000 registered animals (each), and the American Livestock Breeds Conservancy lists 14 as having fewer than 10,000. Preferences for breeds with uniform characteristics and fast growth have pushed heritage (or heirloom) breeds to the margins of the sheep industry. Those that remain are maintained through the efforts of conservation organizations, breed registries, and individual farmers dedicated to their preservation.

Diet

Sheep are exclusively herbivorous mammals. Most breeds prefer to graze on grass and other short roughage, avoiding the taller woody parts of plants that goats readily consume. Both sheep and goats use their lips and tongues to select parts of the plant that are easier to digest or higher in nutrition. Sheep, however, graze well in monoculture pastures where most goats fare poorly. Like all ruminants, sheep have a complex digestive system composed of four chambers, allowing them to break down cellulose from stems, leaves, and seed hulls into simpler carbohydrates. When sheep graze, vegetation is chewed into a mass called a bolus, which is then passed into the first chamber: the rumen. The rumen is a 19 to 38-liter (5 to 10 gal) organ in which feed is fermented via a symbiotic relationship with the bacteria, protozoa, and yeasts of the gut flora. The bolus is periodically regurgitated back to the mouth as cud for additional chewing and salivation. Cud chewing is an adaptation allowing ruminants to graze more quickly in the morning, and then fully chew and digest feed later in the day. This is beneficial as grazing, which requires lowering the head, leaves sheep vulnerable to predators, while cud chewing does not.

During fermentation, the rumen produces gas that must be expelled; disturbances of the organ, such as sudden changes in a sheep's diet, can cause the potentially fatal condition of bloat, when gas becomes trapped in the rumen. After fermentation in the rumen, feed passes in to the reticulum and the omasum; special feeds such as grains may bypass the rumen altogether. After the first three chambers, food moves in to the abomasum for final digestion before processing by the intestines. The abomasum is the only one of the four chambers analogous to the human stomach (being the only one that absorbs nutrients for use as energy), and is sometimes called the "true stomach".

Sheep follow a diurnal pattern of activity, feeding from dawn to dusk, stopping sporadically to rest and chew their cud. Ideal pasture for sheep is not lawn-like grass, but an array of grasses, legumes and forbs. Types of land where sheep are raised vary widely, from pastures that are seeded and improved intentionally to rough, native lands. Common plants toxic to sheep are present in most of the world, and include (but are not limited to) oak and acorns, tomato, yew, rhubarb, potato, and rhododendron.

Sheep are largely grazing herbivores, unlike browsing animals such as goats and deer that prefer taller foliage. With a much narrower face, sheep crop plants very close to the ground and can overgraze a pasture much faster than cattle. For this reason, many shepherds use managed intensive rotational grazing, where a flock is rotated through multiple pastures, giving plants time to recover. Paradoxically, sheep can both cause and solve the spread of invasive plant species. By disturbing the natural state of pasture, sheep and other livestock can pave the way for invasive plants. However, sheep also prefer to eat invasives such as cheatgrass, leafy spurge, kudzu and spotted knapweed over native species such as sagebrush, making grazing sheep effective for conservation grazing. Research conducted in Imperial County, California compared lamb grazing with herbicides for weed control in seedling alfalfa fields. Three trials demonstrated that grazing lambs were just as effective as herbicides in controlling winter weeds.

Entomologists also compared grazing lambs to insecticides for insect control in winter alfalfa. In this trial, lambs provided insect control as effectively as insecticides.

Other than forage, the other staple feed for sheep is hay, often during the winter months. The ability to thrive solely on pasture (even without hay) varies with breed, but all sheep can survive on this diet. Also included in some sheep's diets are minerals, either in a trace mix or in licks.

Naturally, a constant source of potable water is also a fundamental requirement for sheep. The amount of water needed by sheep fluctuates with the season and the type and quality of the food they consume. When sheep feed on large amounts of new growth and there is precipitation (including dew, as sheep are dawn feeders), sheep need less water. When sheep are confined or are eating large amounts of cured hay, more water is typically needed. Sheep also require clean water, and may refuse to drink water that is covered in scum or algae.

Sheep are one of the few livestock animals raised for meat today that have never been widely raised in an intensive, confined animal feeding operation (CAFO). Although there is a growing movement advocating alternative farming styles, a large percentage of beef cattle, pigs, and poultry are still produced under such conditions. In contrast, only some sheep are regularly given high-concentration grain feed, much less kept in confinement. Especially in industrialized countries, sheep producers may fatten market lambs before slaughter (called "finishing") in feedlots. Many sheep breeders flush ewes and rams with a daily ration of grain during breeding to increase fertility. Ewes are also flushed during pregnancy to increase birth weights, as 70% of a lamb's growth occurs in the last five to six weeks of gestation. Otherwise, only lactating ewes and especially old or infirm sheep are commonly provided with grain. Feed provided to sheep must be specially formulated, as most cattle, poultry, pig, and even some goat feeds contain levels of copper that are lethal to sheep. The same danger applies to mineral supplements such as salt licks.

Behaviour and Intelligence

Sheep are prey animals with a strong gregarious instinct, and a majority of sheep behaviors can be understood in these terms. The dominance hierarchy of *Ovis aries* and its natural inclination to follow a leader to new pastures were the pivotal factors in it being one of the first domesticated livestock species. All sheep have a tendency to congregate close to other members of a flock, although this behaviour varies with breed. Farmers exploit this behaviour to keep sheep together on unfenced pastures and to move them more easily. Shepherds may also use herding dogs in this effort, whose highly bred herding ability can assist in moving flocks. Sheep are also extremely food-oriented, and association of humans with regular feeding often results in sheep soliciting people for food. Those who are moving sheep may exploit this behaviour by leading sheep with buckets of feed, rather than forcing their movements with herding.

In regions where sheep have no natural predators, none of the native breeds of sheep exhibit a strong flocking behaviour. Sheep can also become hefted to one particular local pasture (heft) so they do not roam freely in unfenced landscapes. Ewes teach the heft to their lambs, and if whole flocks are culled it must be retaught to the replacement animals. Escaped sheep being led back to pasture with the enticement of food. This method of moving sheep works best with smaller flocks.

Flock dynamics in sheep are, as a rule, only exhibited in a group of four or more sheep. Fewer sheep may not react as normally expected when alone or with few other sheep. For sheep, the primary defense mechanism is simply to flee from danger when their flight zone is crossed. Secondly, cornered sheep may charge or threaten to do so through hoof stamping and aggressive posture. This is particularly true for ewes with newborn lambs.

In displaying flocking, sheep have a strong lead-follow tendency, and a leader often as not is simply the first sheep to move. However, sheep do establish a pecking order through physical displays of dominance. Dominant animals are inclined to be more aggressive with other sheep, and usually feed first at troughs. Primarily among rams, horn size is a factor in the flock hierarchy. Rams with different size horns may be less inclined to fight to establish pecking order, while rams with similarly sized horns are more so. Sheep can become stressed when separated from their flock members. Sheep can recognize individual human and ovine faces, and remember them for years. Relationships in flocks tend to be closest among related sheep: in mixed-breed flocks same-breed subgroups tend to form, and a ewe and her direct descendants often move as a unit within large flocks.

Sheep are frequently thought of as extremely unintelligent animals. A sheep's herd mentality and quickness to flee and panic in the face of stress often make shepherding a difficult endeavor for the uninitiated. Despite these perceptions, a University of Illinois monograph on sheep found them to be just below pigs and on par with cattle in IQ, and some sheep have shown problem-solving abilities; a flock in West Yorkshire, England allegedly found a way to get over cattle grids by rolling on their backs, although documentation of this has relied on anecdotal accounts. In addition to long-term facial recognition of individuals, sheep can also differentiate emotional states through facial characteristics. If worked with patiently, sheep may learn their names, and many sheep are trained to be led by halter for showing and other purposes. Sheep have also responded well to clicker training. Very rarely, sheep are used as pack animals. Tibetan nomads distribute baggage equally throughout a flock as it is herded between living sites.

Reproduction

Sheep follow a similar reproductive strategy to other herd animals. A group of ewes is generally mated by a single ram, who has either been chosen by a breeder or has established dominance through physical contest with other rams (in feral populations). Most sheep are seasonal breeders, although some are able to breed year-round. Ewes generally reach sexual maturity at six to eight months of age, and rams generally at four to six months. Ewes have estrus cycles about every 17 days, during which they emit a scent and indicate readiness through physical displays towards rams. A minority of sheep display a preference for homosexuality (8% on average) or are freemartins (female animals that are behaviorally masculine and lack functioning ovaries).

In feral sheep, rams may fight during the rut to determine which individuals may mate with ewes. Rams, especially unfamiliar ones, will also fight outside the breeding period to establish dominance; rams can kill one another if allowed to mix freely. During the rut, even normally friendly rams may become aggressive towards humans due to increases in their hormone levels.

After mating, sheep have a gestation period of about five months, and normal labour take one to three hours. Although some breeds regularly throw larger litters of lambs, most produce single or twin lambs. During or soon after labour, ewes and lambs may be confined to small lambing jugs, small pens designed to aid both careful observation of ewes and to cement the bond between them and their lambs.

Ovine obstetrics can be problematic. By selectively breeding ewes that produce multiple offspring with higher birth weights for generations, sheep producers have inadvertently caused some domestic sheep to have difficulty lambing; balancing ease of lambing with high productivity is one of the dilemmas of sheep breeding. In the case of any such problems, those present at lambing may assist the ewe by extracting or repositioning lambs. After the birth, ewes ideally break the amniotic sac (if it is not broken during labour), and begin licking clean the lamb. Most lambs will begin standing within an hour of birth. In normal situations, lambs nurse after standing, receiving vital colostrum milk. Lambs that either fail to nurse or that are rejected by the ewe require aid to live, such as bottle-feeding or fostering by another ewe.

After lambs are several weeks old, lamb marking (the process of ear tagging, docking, and castrating) is carried out. Vaccinations are usually carried out at this point as well. Ear tags with numbers are attached, or ear marks are applied for ease of later identification of sheep. Castration is performed on ram lambs not intended for breeding, although some shepherds choose to avoid the procedure for ethical, economic or practical reasons. Ram lambs that will either be slaughtered or separated from ewes before sexual maturity are not usually castrated. Docking, which is the shortening of a lamb's tail, is practised for health reasons. Objections to all these procedures have been raised by animal rights groups, but farmers defend them by saying they solve many practical and veterinary problems, and inflict only temporary pain.

Health

Sheep may fall victim to poisons, infectious diseases, and physical injuries. As a prey species, a sheep's system is adapted to hide the obvious signs of illness, to prevent being targeted by predators. However, there are some obvious signs of ill health, with sick sheep eating little, vocalizing excessively, and being generally listless. Throughout history, much of the money and labour of sheep husbandry has aimed to prevent sheep ailments. Historically, shepherds often created remedies by experimentation on the farm. In some developed countries, including the United States, sheep lack the economic importance for drugs companies to perform expensive clinical trials required to approve drugs for ovine use. In such instances, shepherds resort to illegal, extra-label usage of drugs approved for other animals. In the 20th and 21st centuries, a minority of sheep owners have turned to alternative treatments such as homeopathy, herbalism and even traditional Chinese medicine to treat sheep veterinary problems. Despite some favourable anecdotal evidence, the effectiveness of alternative veterinary medicine has been met with skepticism in scientific journals. The need for traditional anti-parasite drugs and antibiotics is widespread, and is the main impediment to certified organic farming with sheep.

Many breeders take a variety of preventive measures to ward off problems. The first is to ensure that all sheep are healthy when purchased. Many buyers avoid outlets known to be clearing houses for animals culled from healthy flocks as either sick or simply inferior. This can also mean maintaining a closed flock, and quarantining new sheep for a month. Two fundamental preventive programs are maintaining good nutrition and reducing stress in the sheep. Handling sheep in loud, erratic ways causes them to produce cortisol, a stress hormone. This can lead to a weakened immune system, thus making sheep far more vulnerable to disease. Signs of stress in sheep include: excessive panting, teeth grinding, restless movement, wool eating, and wood chewing. Avoiding poisoning is also important; common poisons are pesticide sprays, inorganic fertilizer, motor oil, as well as radiator coolant (the ethylene glycol antifreeze is sweet-tasting).

Common forms of preventive medication for sheep are vaccinations and treatments for parasites. Both external and internal parasites are the most prevalent malady in sheep, and are either fatal, or reduce the productivity of flocks. Worms are the most common internal parasites. They are ingested during grazing, incubate within the sheep, and are expelled through the digestive system (beginning the cycle again). Oral anti-parasitic medicines, known as drenches, are given to a flock to treat worms, sometimes after worm eggs in the feces has been counted to assess infestation levels. Afterwards, sheep may be moved to a new pasture to avoid ingesting the same parasites. External sheep parasites include: lice (for different parts of the body), sheep keds, nose bots, sheep itch mites, and maggots. Keds are blood-sucking parasites that cause general malnutrition and decreased productivity, but are not fatal. Maggots are those of the bot fly and the blow-fly. Fly maggots cause the extremely destructive condition of flystrike. Flies lay their eggs in wounds or wet, manure-soiled wool; when the maggots hatch they burrow into a sheep's flesh, eventually causing death if untreated. In addition to other treatments, crutching (shearing wool from a sheep's rump) is a common preventive method. Nose bots are flies that inhabit a sheep's sinuses, causing breathing difficulties and discomfort. Common signs are a discharge from the nasal passage, sneezing, and frantic movement such as head shaking. External parasites may be controlled through the use of backliners, sprays or immersive sheep dips.

A wide array of bacterial diseases affect sheep. Diseases of the hoof, such as foot rot and foot scald may occur, and are treated with footbaths and other remedies. These painful conditions cause lameness and hinder feeding. Ovine Johne's disease is a wasting disease that affects young sheep. Bluetongue disease is an insect-borne illness causing fever and inflammation of the mucous membranes. Ovine rinderpest (or *peste des petits ruminants*) is a highly contagious and often fatal viral disease affecting sheep and goats.

A few sheep conditions are transmissible to humans. Orf (also known as scabby mouth, contagious ecthyma or soremouth) is a skin disease leaving lesions that is transmitted through skin-to-skin contact. Cutaneous anthrax is also called woolsorter's disease, as the spores can be transmitted in unwashed wool. More seriously, the organisms that can cause spontaneous enzootic abortion in sheep are easily transmitted to pregnant women. Also of concern are the prion disease scrapie and the virus that causes foot-and-mouth disease (FMD), as both can devastate flocks. The latter poses a slight risk to humans. During the 2001 FMD pandemic in the UK, hundreds of sheep were culled and some rare British breeds were at risk of extinction due to this.

Predation

Other than parasites and disease, predation is a threat to sheep and the profitability of sheep raising. Sheep have little ability to defend themselves, compared with other species kept as livestock. Even if sheep survive an attack, they may die from their injuries, or simply from panic. However, the impact of predation varies dramatically with region. In Africa, Australia, the Americas, and parts of Europe and Asia predators are a serious problem. In the United States, for instance, over one third of sheep deaths in 2004 were caused by predation. In contrast, other nations are virtually devoid of sheep predators, particularly islands known for extensive sheep husbandry. Worldwide, canids—including the domestic dog—are responsible for most sheep deaths. Other animals that occasionally prey on sheep include: felines, bears, birds of prey, ravens and feral hogs.

Sheep producers have used a wide variety of measures to combat predation. Pre-modern shepherds used their own presence, livestock guardian dogs, and protective structures such as barns and fencing. Fencing (both regular and electric), penning sheep at night and lambing indoors all continue to be widely used. More modern shepherds used guns, traps, and poisons to kill predators, causing significant decreases in predator populations. In the wake of the environmental and conservation movements, the use of these methods now usually falls under the purview of specially designated government agencies in most developed countries.

The 1970s saw a resurgence in the use of livestock guardian dogs and the development of new methods of predator control by sheep producers, many of them non-lethal. Donkeys and guard llamas have been used since the 1980s in sheep operations, using the same basic principle as livestock guardian dogs. Interspecific pasturing, usually with larger livestock such as cattle or horses, may help to deter predators, even if such species do not actively guard sheep. In addition to animal guardians, contemporary sheep operations may use non-lethal predator deterrents such as motion-activated lights and noisy alarms.

History

Sheep were among the first animals to be domesticated by humankind; sources provide a domestication date between nine and eleven thousand years ago in Mesopotamia. Their wild relatives have several characteristics—such as a relative lack of aggression, a manageable size, early sexual maturity, a social nature, and high reproduction rates—which made them particularly suitable for domestication. Today, *Ovis aries* is an entirely domesticated animal that is largely dependent on man for its health and survival. Feral sheep do exist, but exclusively in areas devoid of large predators (usually islands) and not on the scale of feral horses, goats, pigs, or dogs, although some feral populations have remained isolated long enough to be recognized as distinct breeds.

The exact line of descent between domestic sheep to their wild ancestors is presently unclear. The most common hypothesis states that *Ovis aries* is descended from the Asiatic (*O. orientalis*) species of mouflon. It has been proposed that the European mouflon (*O. musimon*) is an ancient breed of domestic sheep turned feral rather than an ancestor, despite it commonly being cited as ancestor in past literature. A few breeds of sheep, such as the Castlemilk Moorit from Scotland, were formed through crossbreeding with wild European mouflon.

The urial (*O. vignei*) was once thought to have been a forebear of domestic sheep, as they occasionally interbreed with mouflon in the Iranian part of their range. However, the urial, argali (*O. ammon*), and snow sheep (*O. nivicola*) have a different number of chromosomes than other Ovis species, making a direct relationship implausible, and phylogenetic studies show no evidence of urial ancestry. Further studies comparing European and Asian breeds of sheep showed significant genetic differences between the two. Two explanations for this phenomenon have been posited. The first is that there is a currently unknown species or subspecies of wild sheep that contributed to the formation of domestic sheep. A second hypothesis suggests that this variation is the result of multiple waves of capture from wild mouflon, similar to the known development of other livestock.

Initially, sheep were kept solely for meat, milk and skins. Archaeological evidence from statuary found at sites in Iran suggests that selection for woolly sheep may have begun around 6000 BC, but the earliest woven wool garments have only been dated to two to three thousand years later. By that span of the Bronze Age, sheep with all the major features of modern breeds were widespread throughout Western Asia. However, one chief difference between ancient sheep and modern breeds is the technique by which wool could be collected. Primitive sheep cannot be shorn, and must have their wool plucked out by hand in a process called "rooing". This is because fibers called kemps are still longer than the soft fleece. The fleece may also be collected from the field after it falls out. This trait survives today in unrefined breeds such as the Soay and many Shetlands. Indeed, the Soay, along with other Northern European breeds with short tails, unshearable fleece, diminutive size, and horns in both sexes, are closely related to ancient sheep. Originally, weaving and spinning wool was a handicraft practiced at home, rather than an industry. Babylonians, Sumerians, and Persians all depended on sheep; and although linen was the first fabric to be fashioned in to clothing, wool was a prized product. The raising of flocks for their fleece was one of the earliest industries, and flocks were a medium of exchange in barter economies.

In Africa

Sheep entered the African continent not long after their domestication in western Asia. A minority of historians once posited a contentious African theory of origin for *Ovis aries*. This theory is based primarily on rock art interpretations, and osteological evidence from Barbary sheep. The first sheep entered North Africa via Sinai, and were present in ancient Egyptian society between eight and seven thousand years ago. Sheep have always been part of subsistence farming in Africa, but today the only country that keeps an influential number of commercial sheep is South Africa. South African sheep producers, in an attempt to deal with the numerous predators of Africa, invented the livestock protection collar, which holds poison at the jugular to sicken or kill predators.

In Europe

Sheep husbandry spread quickly in Europe. Excavations show that In about 6000 BC, during the Neolithic period of prehistory, the Castelnovien people, living around Chateauneuf-les-Martigues near present-day Marseille in the south of France, were among the first in Europe to keep domestic sheep. Practically from its inception, ancient Greek civilization relied on sheep as primary livestock, and were even said to name individual animals. Scandinavian sheep of a type seen today — with short tails and multi-colored fleece — were also present early on. Later, the Roman Empire kept sheep on a wide scale, and the Romans were an important agent in the spread of sheep raising throughout the continent. Pliny the Elder, in his Natural History (*Naturalis Historia*), speaks at length about sheep and wool. Declaring "Many thanks, too, do we owe to the sheep, both for appeasing the gods, and for giving us the use of its fleece.", he goes on to detail the breeds of ancient sheep and the many colors, lengths and qualities of wool. Romans also pioneered the practice of blanketing sheep, in which a fitted coat (today usually of nylon) is placed over the sheep to improve the cleanliness and luster of its wool.

During the Roman occupation of the British Isles, a large wool processing factory was established in Winchester, England in about 50 AD. By 1000 AD, England and Spain were recognized as the twin centers of sheep production in the Western world. As the original breeders of the fine-wooled merino sheep that have historically dominated the wool trade, the Spanish gained great wealth. Wool money largely financed Spanish rulers and thus the voyages to the New World by conquistadors. The powerful *Mesta* (its full title was *Honrado Concejo de la Mesta*, the Honorable Council of the Mesta) was a corporation of sheep owners mostly drawn from Spain's wealthy merchants, Catholic clergy and nobility that controlled the merino flocks. By the 17th century, the *Mesta* held in upwards of two million head of merino sheep.

Mesta flocks followed a seasonal pattern of transhumance across Spain. In the spring, they left the winter pastures (*invernaderos*) in Extremadura and Andalusia to graze on their summer pastures (*agostaderos*) in Castile, returning again in the autumn. Spanish rulers eager to increase wool profits gave extensive legal rights to the *Mesta*, often to the detriment of local peasantry. The huge merino flocks had a lawful right of way for their migratory routes (*cañadas*). Towns and villages were obliged by law to let the flocks graze on their common land, and the Mesta had its own sheriffs that could summon offending individuals to its own tribunals.

Exportation of merinos without royal permission was also a punishable offense, thus ensuring a near-absolute monopoly on the breed until the mid-18th century. After the breaking of the export ban, fine wool sheep began to be distributed worldwide. The export to Rambouillet by Louis XVI in 1786 formed the basis for the modern Rambouillet (or French Merino) breed. After the Napoleonic Wars and the global distribution of the once-exclusive Spanish stocks of Merinos, sheep raising in Spain reverted to hardy coarse-wooled breeds such as the Churra, and was no longer of international economic significance.

The sheep industry in Spain was an instance of migratory flock management, with large homogenous flocks ranging over the entire nation. The management model used in England was quite different but had a similar importance to economy of the British Empire. Up until the early 20th century, owling (the smuggling of sheep or wool out of the country) was a punishable offense, and to this day the Lord Speaker of the House of Lords sits on a cushion known as the Woolsack.

The high concentration and more sedentary nature of shepherding in the UK allowed sheep especially adapted to their particular purpose and region to be raised, thereby giving rise to an exceptional variety of breeds in relation to the land mass of the country. This greater variety of breeds also produced a valuable variety of products to compete with the superfine wool of Spanish sheep. By the time of Elizabeth I's rule, sheep and wool trade was the primary source of tax revenue to the Crown of England and the country was a major influence in the development and spread of sheep husbandry.

An important event not only in the history of domestic sheep, but of all livestock, was the work of Robert Bakewell in the 18th century. Before his time, breeding for desirable traits was often based on chance, with no scientific process for selection of breeding stock.

Bakewell established the principles of selective breeding—especially line breeding—in his work with sheep, horses and cattle; his work later influenced Gregor Mendel and Charles Darwin. His most important contribution to sheep was the development of the Leicester Longwool, a quick-maturing breed of blocky conformation that formed the basis for many vital modern breeds. Today, the sheep industry in the UK has diminished significantly, though pedigreed rams can still fetch around 100,000 Pounds sterling at auction.

In the Americas

No ovine species native to the Americas has ever been domesticated, despite being closer genetically to domestic sheep than many Asian and European species. The first domestic sheep in North America—most likely of the Churra breed—arrived with Christopher Columbus' second voyage in 1493. The next transatlantic shipment to arrive was with Hernán Cortés in 1519, landing in Mexico. No export of wool or animals is known to have occurred from these populations, but flocks did disseminate throughout what is now Mexico and the Southwest United States with Spanish colonists. Churras were also introduced to the Navajo tribe of Native Americans, and became a key part of their livelihood and culture. The modern presence of the Navajo-Churro breed is a result of this heritage.

North America

The next transport of sheep to North America was not until 1607, with the voyage of the *Susan Constant* to Virginia. However, the sheep that arrived in that year were all slaughtered because of a famine, and a permanent flock was not to reach the colony until two years later in 1609. In two decades time, the colonists had expanded their flock to a total of 400 head. By the 1640s there were about 100,000 head of sheep in the 13 colonies, and in 1662, a woolen mill was built in Watertown, Massachusetts. Especially during the periods of political unrest and civil war in Britain spanning the 1640s and 50s which disrupted maritime trade, the colonists found it pressing to produce wool for clothing.

Many islands off the coast were cleared of predators and set aside for sheep: Nantucket, Long Island, Martha's Vineyard and small islands in Boston Harbor were notable examples. There remain some rare breeds of American sheep—such as the Hog Island sheep—that were the result of island flocks. Placing semi-feral sheep and goats on islands was common practice in colonization during this period. Early on, the British government banned further export of sheep to the Americas, or wool from it, in an attempt to stifle any threat to the wool trade in the British Isles. One of many restrictive trade measures that precipitated the American Revolution, the sheep industry in the Northeast grew despite the bans.

Gradually, beginning in the 19th century, sheep production in the U.S. moved westward. Today, the vast majority of flocks reside on Western range lands. During this westward migration of the industry, competition between sheep (sometime called "range maggots") and cattle operations grew more heated, eventually erupting into range wars. Other than simple competition for grazing and water rights, cattlemen believed that the secretions of the foot glands of sheep made cattle unwilling to graze on places where sheep had stepped. As sheep production centered on the U.S. western ranges, it became associated with other parts of Western culture, such as the rodeo. Another effect of the westward movement of sheep flocks in North America was the decline of wild species such as Bighorn sheep (*O. canadesis*). Most diseases of domestic sheep are transmittable to wild ovines, and such diseases, along with overgrazing and habitat loss, are named as primary factors in the plummeting numbers of wild sheep. Sheep production peaked in North America during 1940s and 50s at more than 55 million head. Henceforth and continuing today, the number of sheep in North America has steadily declined with wool prices and the lessening American demand for sheep meat.

South America

In South America, especially in Patagonia, there is an active modern sheep industry. Sheep keeping was largely introduced through immigration to the continent by Spanish and British peoples, for whom sheep were a major industry during the period. South America has a large number of sheep, but the highest-producing nation (Brazil) kept only just over 15 million head in 2004, far fewer than most centers of sheep husbandry. The primary challenges to the sheep industry in South America are the phenomenal drop in wool prices in the late 20th century and the loss of habitat through logging and overgrazing. The most influential region internationally is that of Patagonia, which has been the first to rebound from the fall in wool prices. With few predators and almost no grazing competition (the only large native grazing mammal is the guanaco), the region is prime land for sheep raising. The most exceptional area of production is surrounding the La Plata river in the Pampas region. Sheep production in Patagonia peaked in 1952 at more than 21 million head, but has steadily fallen to fewer than ten today. Most operations focus on wool production for export from Merino and Corriedale sheep; the economic sustainability of wool flocks has fallen with the drop in prices, while the cattle industry continues to grow.

In Australia and New Zealand

Australia and New Zealand are crucial players in the contemporary sheep industry, and sheep are an iconic part of both countries' culture and economy. New Zealand has the highest density of sheep per capita (sheep outnumber the human population 12 to 1), and Australia is the world's indisputably largest exporter of sheep and cattle. In 2007, New Zealand even declared 15 February their official National Lamb Day to celebrate the country's history of sheep production.

The First Fleet brought the initial population of 70 sheep from the Cape of Good Hope to Australia in 1788. The next shipment was of 30 sheep from Calcutta and Ireland in 1793. All of the early sheep brought to Australia were exclusively used for the dietary needs of the penal colonies. The beginnings of the Australian wool industry were due to the efforts of Captain John Macarthur. At Macarthur's urging 16 Spanish merinos were imported in 1797, effectively beginning the Australian sheep industry. By 1801 Macarthur had 1,000 head of sheep, and in 1803 he exported 111 kilograms (245 lb) of wool to England. Today, Macarthur is generally thought of as the father of the Australian sheep industry.

The growth of the sheep industry in Australia was explosive. In 1820, the continent held 100,000 sheep, a decade later it had one million. By 1840, New South Wales alone kept 4 million sheep; flock numbers grew to 13 million in a decade. While much of the growth in both nations was due to the active support of Britain in its desire for wool, both worked independently to develop new high-production breeds: the Corriedale, Coolalee, Coopworth, Perendale, Polwarth, Booroola Merino, Peppin Merino, and Poll Merino were all created in New Zealand or Australia. Wool production was a fitting industry for colonies far from their home nations. Before the advent of fast air and maritime shipping, wool was one of the few viable products that was not subject to spoiling on the long passage back to British ports. The abundant new land and milder winter weather of the region also aided the growth of the Australian and New Zealand sheep industries.

Flocks in Australia have always been largely range bands on fenced land, and are aimed at production of medium to superfine wool for clothing and other products as well as meat. New Zealand flocks are kept in a fashion similar to English ones, in fenced holdings without shepherds. Although wool was once the primary income source for New Zealand sheep owners (especially during the New Zealand wool boom), today it has shifted to meat production for export.

Animal Welfare Concerns

The Australian sheep industry is the only sector of the industry to receive international criticism for its practices. Sheep stations in Australia are cited in *Animal Liberation*, the seminal book of the animal rights movement, as the author's primary evidence in his argument against retaining sheep as a part of animal agriculture. The practice of mulesing, in which skin is cut away from an animal's perineal area to prevent cases of the fatal condition flystrike, has been condemned by PETA as being painful and unnecessary. In response, a program of phasing out mulesing is currently being implemented, and some mulesing operations are being carried out with the use of anaesthetic. The Animal Welfare Advisory Committee to the New Zealand Ministry of Agriculture *Code of recommendations and minimum standards for the welfare of Sheep*, considers mulesing a "special technique" which is performed on some Merino sheep at a small number of farms in New Zealand.

Most of the sheep meat exported from Australia is either frozen carcases to the UK or is live export to the Middle East. Shipped on livestock carriers in what has been called crowded, unsafe conditions by critics, live sheep are desired by Middle Eastern nations to meet the requirements of ritual halal slaughter. Opponents of the export—

such as PETA—say that sheep exported to countries outside the jurisdiction of Australia's animal cruelty laws are treated with horrendous brutality and that halal facilities exist in Australia to make export of live animals redundant. A few celebrities and companies have pledged to boycott all Australian sheep products in protest.

Economic Importance

Sheep are an important part of the global agricultural economy. However, their once-vital status has been largely replaced by other livestock species, especially the pig, chicken, and cow. China, Australia, India, and Iran have the largest modern flocks, and serve both local and exportation needs for wool and mutton. Other countries such as New Zealand have smaller flocks but retain a large international economic impact due to their export of sheep products. Sheep also play a major role in many local economies, which may be niche markets focused on organic or sustainable agriculture and local food customers. Especially in developing countries, such flocks may be a part of subsistence agriculture rather than a system of trade. Sheep themselves may be a medium of trade in barter economies.

Domestic sheep provide a wide array of raw materials. Wool was one of the first textiles, although in the late 20th century wool prices began to fall dramatically as the result of the popularity and cheap prices for synthetic fabrics. For many sheep owners, the cost of shearing is greater than the possible profit from the fleece, making subsisting on wool production alone practically impossible without farm subsidies. Fleeces are used as material in making alternative products such as wool insulation. In the 21st century, the sale of meat is the most profitable enterprise in the sheep industry, even though far less sheep meat is consumed than chicken, pork or beef.

Sheepskin is likewise used for making clothes, footwear, rugs, and other products. Byproducts from the slaughter of sheep are also of value: sheep tallow can be used in candle and soap making, sheep bone and cartilage has been used to furnish carved items such as dice and buttons as well as rendered glue and gelatin. Sheep intestine can be formed into sausage casings, and lamb intestine has been formed into surgical sutures, as well as strings for musical instruments and tennis rackets. Sheep droppings, which are high in cellulose, have even been sterilized and mixed with traditional pulp materials to make paper. Of all sheep byproducts, perhaps the most valuable is lanolin: the water-proof, fatty substance found naturally in sheep's wool and used as a base for innumerable cosmetics and other products.

Some farmers who keep sheep also make a profit from live sheep. Providing lambs for youth programs such as 4-H and competition at agricultural shows is often a dependable avenue for the sale of sheep. Farmers may also choose to focus on a particular breed of sheep in order to sell registered purebred animals, as well as provide a ram rental service for breeding. The most valuable sheep ever sold to date was a purebred Texel ram that fetched £231,000 at auction. The previous record holder was a Merino ram sold for £205,000 in 1989. A new option for deriving profit from live sheep is the rental of flocks for grazing; these "mowing services" are hired in order to keep unwanted vegetation down in public spaces and to lessen fire hazard.

Despite the falling demand and price for sheep products in many markets, sheep have distinct economic advantages when compared with other livestock. They do not require the expensive housing, such as that used in the intensive farming of chickens or pigs. They are an efficient use of land; roughly six sheep can be kept on the amount that would suffice for a single cow or horse. Sheep can also consume plants, such as noxious weeds, that most other animals will not touch, and produce more young at a faster rate. Also, in contrast to most livestock species, the cost of raising sheep is not necessarily tied to the price of feed crops such as grain, soybeans and corn. Combined with the lower cost of quality sheep, all these factors combine to equal a lower overhead for sheep producers, thus entailing a higher profitability potential for the small farmer. Sheep are especially beneficial for independent producers, including family farms with limited resources, as the sheep industry is one of the few types of animal agriculture that has not been vertically integrated by agribusiness.

As Food

Sheep meat and milk were one of the earliest staple proteins consumed by human civilization after the transition from hunting and gathering to agriculture. Sheep meat prepared for food is known as either mutton or lamb. "Mutton" is derived from the Old French *moton*, which was the word for sheep used by the Anglo-Norman rulers of much of the British Isles in the Middle Ages. This became the name for sheep meat in English, while the Old English word *sceap* was kept for the live animal. Throughout modern history, "mutton" has been limited to the meat of mature sheep usually at least two years of age; "lamb" is used for that of immature sheep less than a year.

In the 21st century, the nations with the highest consumption of sheep meat are the Persian Gulf states, New Zealand, Australia, Greece, Uruguay, the United Kingdom and Ireland. These countries eat 14–

40 lbs (3–18 kg) of sheep meat per capita, per annum. Sheep meat is also popular in France, Africa (especially the Maghreb), the Caribbean, the rest of the Middle East, India, and parts of China. This often reflects a past history of sheep production. In these countries in particular, dishes comprising alternative cuts and offal may be popular or traditional. Sheep testicles—called animelles or lamb fries—are considered a delicacy in many parts of the world. Perhaps the most unusual dish of sheep meat is the Scottish haggis, composed of various sheep innards cooked along with oatmeal and chopped onions inside its stomach. In comparison, countries such as the U.S. consume only a pound or less (under 0.5 kg), with Americans eating 50 pounds (22 kg) of pork and 65 pounds (29 kg) of beef. In addition, such countries rarely eat mutton, and may favour the more expensive cuts of lamb: mostly lamb chops and leg of lamb.

Though sheep's milk may be drunk rarely in fresh form, today it is used predominantly in cheese and yogurt making. Sheep have only two teats, and produce a far smaller volume of milk than cows. However, as sheep's milk contains far more fat, solids, and minerals than cow's milk, it is ideal for the cheese-making process. It also resists contamination during cooling better because of its much higher calcium content. Well-known cheeses made from sheep milk include the Feta of Bulgaria and Greece, Roquefort of France, Manchego from Spain, the Pecorino Romano (the Italian word for sheep is *pecore*) and Ricotta of Italy. Yogurts, especially some forms of strained yogurt, may also be made from sheep milk. Many of these products are now often made with cow's milk, especially when produced outside their country of origin. Sheep milk contains 4.8% lactose, which may affect those who are intolerant.

In Science

Sheep are generally too large and reproduce too slowly to make ideal research subjects, and thus are not a common model organism. They have, however, played an influential role in some fields of science. In particular, the Roslin Institute of Edinburgh, Scotland used sheep for genetics research that produced groundbreaking results. In 1995, two ewes named Megan and Morag were the first mammals cloned from differentiated cells. A year later, a Finnish Dorset sheep named Dolly, dubbed "the world's most famous sheep" in *Scientific American*, was the first mammal to be cloned from an adult somatic cell. Following this, Polly and Molly were the first mammals to be simultaneously cloned and transgenic. As of 2008, the sheep genome has not been fully sequenced, although a detailed genetic map has been published, and a draft version of the complete genome produced by assembling sheep DNA sequences using information given by the genomes of other mammals.

In the study of natural selection, the population of Soay sheep that remain on the island of Hirta have been used to explore the relation of body size and coloration to reproductive success. Soay sheep come in several colors, and researchers investigated why the larger, darker sheep were in decline; this occurrence contradicted the rule of thumb that larger members of a population tend to be more successful reproductively. The feral Soays on Hirta are especially useful subjects because they are isolated.

Sheep are one of the few animals where the molecular basis of the diversity of male sexual preferences has been examined. However, this research has been controversial, and much publicity has been produced by a study at the Oregon Health and Science University that investigated the mechanisms that produce homosexuality in rams. Organizations such as PETA campaigned against the study, accusing scientists of trying to cure homosexuality in the sheep. OHSU and the involved scientists vehemently denied such accusations.

Domestic sheep are sometimes used in medical research, particularly for researching cardiovascular physiology, in areas such as hypertension and heart failure. Pregnant sheep are also a useful model for human pregnancy, and have been used to investigate the effects on fetal development of malnutrition and hypoxia. In behavioral sciences, sheep have been used in isolated cases for the study of facial recognition, as their mental process of recognition is qualitatively similar to humans.

Cultural Impact

Sheep have had a strong presence in many cultures, especially in areas where they form the most common type of livestock. In the English language, to call someone a sheep or ovine may allude that they are timid and easily led, if not outright stupid. In contradiction to this image, male sheep are often used as symbols of virility and power, such as for the St. Louis Rams and the Dodge Ram. Sheep are key symbols in fables and nursery rhymes like *The Wolf in Sheep's Clothing*, *Little Bo Peep*, *Baa, Baa, Black Sheep*, and *Mary Had a Little Lamb*. Novels such as George Orwell's *Animal Farm*, Haruki Murakami's *A Wild Sheep Chase*, Thomas Hardy's *Far from the Madding Crowd* and *Three Bags Full: A Sheep Detective Story* utilize sheep as characters or plot devices. Poems like William Blake's "The Lamb", songs such as Pink Floyd's *Sheep* and Bach's aria *Sheep may safely graze* (*Schafe können sicher weiden*) use sheep for metaphorical purposes. In more recent popular culture, the 2007 film *Black Sheep* exploits sheep for horror and comedic effect, ironically turning them into blood-thirsty killers. Counting sheep is popularly said to be an aid to sleep, and some ancient systems of counting sheep persist today. Sheep also enter in colloquial sayings and idiom frequently with such phrases as "black sheep". To call an individual a black sheep implies that they are an odd or disreputable member of a group. This usage derives from the recessive trait that causes an occasional black lamb to be born in to an entirely white flock. These black sheep were considered undesirable by shepherds, as black wool is not as commercially viable as white wool. Citizens who accept overbearing governments have been referred to by the Portmanteau neologism of sheeple. Somewhat differently, the adjective "sheepish" is also used to describe embarrassment.

In religion and folklore

In antiquity, symbolism involving sheep cropped up in religions in the ancient Near East, the Mideast, and the Mediterranean area: Çatalhöyük, ancient Egyptian religion, the Cana'anite and Phoenician tradition, Judaism, Greek religion, and others. Religious symbolism and ritual involving sheep began with some of the first known faiths: skulls of rams (along with bulls) occupied central placement in shrines at the Çatalhöyük settlement in 8,000 BCE. In Ancient Egyptian religion, the ram was the symbol of several gods: Khnum, Heryshaf and Amun (in his incarnation as a god of fertility). Other deities occasionally shown with ram features include: the goddess Ishtar, the Phoenician god Baal-Hamon, and the Babylonian god Ea-Oannes. In Madagascar, sheep were not eaten as they were believed to be incarnations of the souls of ancestors.

There are also many ancient Greek references to sheep: that of Chrysomallos, the golden-fleeced ram, continuing to be told through into the modern era. Astrologically, *Aries*, the ram, is the first sign of the classical Greek zodiac and the sheep is also the eighth of the twelve animals associated with the 12-year cycle of in the Chinese zodiac, related to the Chinese calendar. In Mongolia, shagai are an ancient form of dice made from the cuboid bones of sheep that are often used for fortunetelling purposes.

Sheep play an important role in all the Abrahamic faiths; Abraham, Isaac, Jacob, Moses, King David and the Islamic prophet Muhammad were all shepherds. According to the Biblical story of the Binding of Isaac, a ram is sacrificed as a substitute for Isaac after an angel stays Abraham's hand (in the Islamic tradition, Abraham was about to sacrifice Ishmael). Eid al-Adha is a major annual festival in Islam in which sheep (or other animals) are sacrificed in remembrance of this act. Sheep are also occasionally sacrificed to commemorate important secular events in Islamic cultures. Greeks and Romans also sacrificed sheep regularly in religious practice, and Judaism also once sacrificed sheep as a Korban (sacrifice), such as the Passover lamb. Ovine symbols—such as the ceremonial blowing of a shofar—still find a presence in modern Judaic traditions. Followers of Christianity are collectively often referred to as a flock, with Christ as the Good Shepherd, and sheep are an element in the Christian iconography of the birth of Jesus. Some Christian saints are considered patrons of shepherds, and even of sheep themselves. Christ is also portrayed as the Sacrificial lamb of God (*Agnus Dei*) and Easter celebrations in Greece and Romania traditionally feature a meal of Paschal lamb.

Goat

The domestic goat (*Capra aegagrus hircus*) is a subspecies of goat domesticated from the wild goat of southwest Asia and Eastern Europe. The goat is a member of the Bovidae family and is closely related to the sheep as both are in the goat-antelope subfamily Caprinae. There are over three hundred distinct breeds of goat.

Goats are one of the oldest domesticated species. Goats have been used for their milk, meat, hair, and skins over much of the world. In the twentieth century they also gained in popularity as pets.

Female goats are referred to as *does* or *nannies*, intact males as *bucks* or *billies*; their offspring are *kids*. Note that many goat breeders prefer the terms "buck" and "doe" to "billy" and "nanny". Castrated males are *wethers*. Goat meat from younger animals is called *kid* or *cabrito*, and from older animals is sometimes called *chevon*, or in some areas "mutton".

Etymology

The Modern English word *goat* comes from the Old English *gât* which meant "she-goat", and this in turn derived from Proto-Germanic **gaitaz* (cf. Old Norse and Dutch *geit* "goat", German *Geiß* "she-goat", and Gothic *gaits* "goat"), ultimately from Proto-Indo-European **ghaidos* meaning "young goat" (cf. Latin *haedus* "kid"), itself perhaps from a root meaning "jump" (assuming that Old Church Slavonic *zajêcĐ* "hare", Sanskrit *jihîte* "he moves" are related). To refer to the male of the species, Old English used *bucca* (which survives as "buck") until a shift to *he-goat* (and *she-goat*) occurred in the late 12th century. "Nanny goat" originated in the 18th century and "billy goat" in the 19th.

History

Goats are among the earliest animals domesticated by humans. The most recent genetic analysis confirms the archaeological evidence that the Anatolian Zagros are the likely origin of almost all domestic goats today. Another major genetic source of modern goats is the Bezoar goat, distributed from the mountainous regions of Asia Minor across the Middle East to Sind.

Neolithic farmers began to keep goats for access to milk and meat, primarily, as well as for their dung, which was used as fuel and their bones, hair, and sinew for clothing, building, and tools. The earliest remnants of domesticated goats dating 10,000 years before present are found in Ganj Dareh in Iran. Goat remains have been found at archaeological sites in Jericho, Choga, Mami, Djeitun and Cayonu, dating the domestication of goats in western Asia at between 8000 and 9000 years ago. Historically, goat hide has been used for water and wine bottles in both traveling and transporting wine for sale. It has also been used to produce parchment.

Anatomy and Health

Most goats naturally have two horns, of various shapes and sizes depending on the breed. All goats have horns unless they are "polled" meaning they have one parent with a dominant polled gene. There have been incidents of polycerate goats (having as many as eight horns), although this is a genetic rarity thought to be inherited. Their horns are made of living bone surrounded by keratin and other proteins, and are used for defense, dominance, and territoriality.

Goats are ruminants. They have a four-chambered stomach consisting of the rumen, the reticulum, the omasum, and the abomasum. As with other mammal ruminants, they are even-toed ungulates. The females have an udder consisting of two teats, in contrast to cattle, which have four teats.

Goats have horizontal slit-shaped pupils, an adaptation which increases peripheral depth perception. Because goats' irises are usually pale, the pupils are much more visible than in animals with horizontal pupils, but very dark irises, such as cattle, deer, most horses and many sheep. Both male and female goats have beards, and many types of goat (most commonly dairy goats, dairy-cross boers, and pygmy goats) may have wattles, one dangling from each side of the neck.

Some breeds of sheep and goats look similar, but they can usually be told apart because goat tails are short and point up, whereas sheep tails hang down and are usually longer and bigger – though some (like those of Northern European short-tailed sheep) are short, and longer ones are often docked.

Reproduction

Goats reach puberty between 3 and 15 months of age, depending on breed and nutrition status. Many breeders prefer to postpone breeding until the doe has reached 70% of the adult weight. However, this separation is rarely possible in extensively managed, open range herds. In temperate climates and among the Swiss breeds, the breeding season commences as the day length shortens, and ends in early spring or before. In equatorial regions, goats are able to breed at any time of the year. Successful breeding in these regions depends more on available forage than on day length. Does of any breed or region come into heat every 21 days for 2 to 48 hours. A doe in heat typically flags (vigorously wags) her tail often, stays near the buck if one is present, becomes more vocal, and may also show a decrease in appetite and milk production for the duration of the heat.

Bucks (intact males) of Swiss and northern breeds come into rut in the fall as with the doe's heat cycles. Bucks of equatorial breeds may show seasonal reduced fertility but, as with the does, are capable of breeding at all times. Rut is characterized by a decrease in appetite and obsessive interest in the does. A buck in rut will display flehmen lip curling and will urinate on his forelegs and face. Sebaceous scent glands at the base of the horns add to the male goat's odour, which is important to make him attractive to the female. Some does will not mate with a buck which has been de-scented.

In addition to natural mating, artificial insemination has gained popularity among goat breeders, as it allows easy access to a wide variety of bloodlines.

Gestation length is approximately 150 days. Twins are the usual result, with single and triplet births also common. Less frequent are litters of quadruplet, quintuplet, and even sextuplet kids. Birthing, known as *kidding*, generally occurs uneventfully. Just before kidding, the doe will have a sunken area around the tail and hip, as well as heavy breathing. She may have a worried look, become restless and display great affection for her keeper. The mother often eats the placenta, which gives her much needed nutrients, helps stanch her bleeding, and parallels the behaviour of wild herbivores such as deer to reduce the lure of the birth scent for predators.

Freshening (coming into milk production) occurs at kidding. Milk production varies with the breed, age, quality, and diet of the doe;

dairy goats generally produce between 660 to 1,800 L (1,500 and 4,000 lb) of milk per 305 day lactation. On average, a good quality dairy doe will give at least 6 lb (2.7 l) of milk per day while she is in milk. A first time milker may produce less, or as much as 16 lb (7.3 l), or more of milk in exceptional cases. After the 305 day lactation, the doe will "dry off", typically after she has been bred. Occasionally, goats that have not been bred and are continuously milked will continue lactation beyond the typical 305 days. Meat, fibre, and pet breeds are not usually milked and simply produce enough for the kids until weaning. Western European-origin goats without horns (polled) frequently produce intersex offspring. These are generally female animals with male characteristics, and are infertile.

Diet

Goats are reputed to be willing to eat almost anything, except tin cans and cardboard boxes. While goats will not actually eat inedible material, they are browsing animals, not grazers like cattle and sheep, and (coupled with their natural curiosity) will chew on and taste just about anything resembling plant matter in order to decide whether it is good to eat, including cardboard and paper labels from tin cans. Another possibility is that the goats are curious about the unusual smells of leftover food in discarded cans or boxes.

Aside from sampling many things, goats are quite particular in what they actually consume, preferring to browse on the tips of woody shrubs and trees, as well as the occasional broad-leaved plant. However, it can fairly be said that their plant diet is extremely varied, and includes some species which are otherwise toxic. They will seldom consume soiled food or contaminated water unless facing starvation. This is one reason goat rearing is most often free ranging, since stall-fed goat rearing involves extensive upkeep and is seldom commercially viable. Goats prefer to browse on shrubbery and weeds, more like deer than sheep, preferring them to grasses. Nightshade is poisonous; wilted fruit tree leaves can also kill goats. Silage (corn stalks) is not good for goats, but haylage can be used if consumed immediately after opening. Alfalfa is their favorite hay; fescue is the least palatable and least nutritious. Mold in a goat's feed can make it sick and possibly kill it. Goats should not be fed grass showing any signs of mold.

The digestive physiology of a very young kid (like the young of other ruminants) is essentially the same as that of a monogastric animal. Milk digestion begins in the abomasum, the milk having bypassed the rumen via closure of the reticular/esophageal groove during suckling. At birth, the rumen is undeveloped, but as the kid begins to consume solid feed, the rumen soon increases in size and in its capacity to absorb nutrients.

Behaviour

Goats are extremely curious and intelligent. They are easily trained to pull carts and walk on leads. Ches McCartney, nicknamed "the goat man", toured the United States for over three decades in a wagon pulled by a herd of pet goats. They are also known for escaping their pens. Goats will test fences, either intentionally or simply because they are handy to climb on. If any of the fencing can be spread, pushed over or down, or otherwise be overcome, the goats will escape. Being very intelligent, once a weakness in the fence has been discovered, it will be exploited repeatedly. Goats are very coordinated and can climb and hold their balance in the most precarious places. Goats are also widely known for their ability to climb trees, although the tree generally has to be on somewhat of an angle. The vocalization goats make is called bleating. Goats have an intensely inquisitive and intelligent nature: they will explore anything new or unfamiliar in their surroundings. They do so primarily with their prehensile upper lip and tongue. This is why they investigate items such as buttons, camera cases or clothing (and many other things besides) by nibbling at them, occasionally even eating them. When handled as a group, goats tend to display less clumping behaviour than sheep, and when grazing undisturbed, tend to spread across the field or range, rather than feed side-by-side as do sheep. When nursing young, goats will leave their kids separated ('lying out') rather than clumped as do sheep. They will generally turn and face an intruder and bucks are more likely to charge or butt at humans than are rams.

Diseases

While goats are generally considered hardy animals and in many situations receive little medical care, they are subject to a number of diseases. Among the conditions affecting goats are respiratory diseases, including pneumonia, foot rot, internal parasites, pregnancy toxosis and feed toxicity. Goats can become infected with various viral and bacterial diseases such as foot-and-mouth disease, caprine arthritis encephalitis, caseous lymphadenitis, pinkeye, mastitis, and pseudorabies. They can transmit a number of zoonotic diseases to people, such as tuberculosis, brucellosis, Q-fever, and rabies.

Life Expectancy

Life expectancy for goats is between 15 and 18 years. An instance of a goat reaching the age of 24 has been reported.

Several factors can reduce this average expectancy, however; problems during kidding can lower a doe's expected life span to 10 or 11, and stresses of going into rut can lower a buck's expected life span to 8 or 10.

Goats in Agriculture

Goat husbandry is common through the Norte Chico region in Chile, but also produces severe erosion and desertification.

Image from upper Limarí River

A goat is useful to humans either living or dead, first as a renewable provider of milk, manure, and fibre, and then as meat and hide. Some charities provide goats to impoverished people in poor countries, because goats are easier and cheaper to manage than cattle, and have multiple uses. In addition, goats are used for driving and packing purposes. For instance, the intestine is used to make "catgut", which is still in use as a material for internal human surgical sutures and strings for musical instruments. The horn of the goat, which signifies wellbeing (Cornucopia), is also used to make spoons.

Husbandry

Husbandry, or animal care and use, varies from region to region and from culture to culture. The particular housing used for goats depends not only on the intended use of the goat but also on the region of the world where they are raised. Historically, domestic goats were generally kept in herds that wandered on hills or other grazing areas, often tended by goatherds who were frequently children or adolescents, similar to the more widely known shepherd. These methods of herding are still used today.

In some parts of the world, especially Europe and North America, distinct breeds of goat are kept for dairy (milk) and for meat production. As with cattle, only the females give milk. Excess male kids of dairy breeds are typically slaughtered for meat. Both does and bucks of meat breeds may be slaughtered for meat, as well as older animals of any breed. The meat of older bucks (more than 1 year old) is generally considered not desirable for meat for human consumption. Castration at a young age prevents the development of typical buck odour.

Dairy goats are generally pastured in summer and may be stabled during the winter. As dairy does are milked daily, they are generally kept close to the milking shed. Their grazing is typically supplemented with hay and with concentrates. Stabled goats may be kept in stalls similar to horses, or in larger group pens. In the US system, does are generally re-bred annually. In some European commercial dairy systems, the does are bred only twice, and are milked continuously for several years after the second kidding.

Meat goats are more frequently pastured year-round, and may be kept many miles from barns. Angora and other fibre breeds are also kept on pasture or range. Range-kept and pastured goats may be supplemented with hay or concentrates, but this happens most frequently during the winter or dry seasons.

In India, Nepal, and much of Asia, goats are kept largely for milk production, both in commercial and household settings. The goats in this area may be kept closely housed or may be allowed to range for fodder. The Salem Black goat is herded to pasture in fields and along roads during the day but is kept penned at night for safe-keeping.

In Africa and the Mideast, goats are typically run in flocks with sheep. This maximizes the production per acre, as goats and sheep prefer different food plants. Multiple types of goat-raising are found in Ethiopia, where four main types of goat raising have been identified: goats kept pastured in annual crop systems, goats kept in perennial crop systems, goats kept with cattle, and goats kept in arid areas under pastoral (nomadic) herding systems.

In all four systems, however, goats were typically kept in extensive systems, with few purchased inputs. Household goats are traditionally kept in Nigeria. While many goats are allowed to wander the homestead or village, others are kept penned and fed in what is called a 'cut-and-carry' system. This type of husbandry is also used in parts of Latin America. Cut-and-carry, which refers to the practice of cutting down grasses, corn or cane for feed rather than allowing the animal access to the field, is particularly suited for types of feed, such as corn or cane, that are easily destroyed by trampling.

Pet goats may be found in many parts of the world when a family keeps one or more animals for emotional reasons rather than as production animals. It is becoming more common for goats to be kept exclusively as pets in North America and Europe.

Meat

The taste of goat kid meat is similar to that of spring lamb meat; in fact, in the English-speaking islands of the Caribbean, and in some parts of Asia, particularly Pakistan and India, the word "mutton" is used to describe both goat and lamb meat. However, some compare the taste of goat meat to veal or venison, depending on the age and condition of the goat. Its flavor is said to be primarily linked to the presence of 4-methyloctanoic and 4-methylnonanoic acid. It can be prepared in a variety of ways including stewed, baked, grilled, barbecued, minced, canned, fried, curried, or made into sausage. Due to its low fat content, the meat can toughen at high temperatures without additional moisture. One of the most popular goats grown for meat is the South African Boer, introduced into the United States in the early 1990s. The New Zealand Kiko is also considered a meat breed, as is the myotonic or "fainting goat", a breed originating in Tennessee.

Milk, Butter and Cheese

Goats produce approximately 2% of the world's total annual milk supply. Some goats are bred specifically for milk. If the strong-smelling buck is not separated from the does, his scent will affect the milk.

Doe milk naturally has small, well-emulsified fat globules, which means the cream remains suspended in the milk, instead of rising to the top, as in raw cow milk; therefore, it does not need to be homogenized. Indeed, if the milk is going to be used to make cheese it is recommended that it is not homogenized as this changes the structure of the milk impacting the culture's ability to coagulate the milk and the final quality and yield of cheese.

Dairy goats in their prime, which is generally around the third or fourth lactation cycle, average 6 to 8 pounds (2.7 to 3.6 kg) of milk production daily (roughly 3 to 4 US quarts (2.7 to 3.6 liters)) during a ten-month lactation, producing more just after freshening and gradually dropping in production toward the end of their lactation. The milk generally averages 3.5 percent butterfat. A doe may be expected to reach her heaviest production during her third or fourth lactation.

Doe milk is commonly processed into cheese, butter, ice cream, yoghurt, cajeta and other products. Goat cheese is known as *chèvre* in France, after the French word for "goat". Some varieties include Rocamadour and Montrachet. Goat butter is white because goats produce milk with the yellow beta-carotene converted to a colorless form of vitamin A.

Nutrition

The American Academy of Pediatrics discourages feeding infants milk derived from goats. An April 2010 case report summarizes their recommendation and presents "a comprehensive review of the consequences associated with this dangerous practice," also stating, "Many infants are exclusively fed unmodified goat's milk as a result of cultural beliefs as well as exposure to false online information.

Anecdotal reports have described a host of morbidities associated with that practice, including severe electrolyte abnormalities, metabolic acidosis, megaloblastic anemia, allergic reactions including life-threatening anaphylactic shock, hemolytic uremic syndrome, and infections." Untreated caprine brucellosis results in a 2% case fatality rate. According to the United States Department of Agriculture (USDA), doe milk is not recommended for human infants because it contains "inadequate quantities of iron, folate, vitamins C and D, thiamin, niacin, vitamin B6, and pantothenic acid to meet an infant's nutritional needs" and may cause harm to an infant's kidneys and could cause metabolic damage.

The Department of Health in the United Kingdom has repeatedly released statements stating on various occasions that "Goats' milk is not suitable for babies, and infant formulas and follow-on formulas based on goats' milk protein have not been approved for use in Europe," and "infant milks based on goats' milk protein are not suitable as a source of nutrition for infants.".

On the other hand, some farming groups promote the practice. For example Small Farm Today in 2005 claimed beneficial use in invalid and convalescent diets, proposing that glycerol ethers, possibly important in nutrition for nursing infants, are much higher in doe milk than in cow milk. A 1970 book on animal breeding claimed that doe milk differs from cow or human milk by having higher digestibility, distinct alkalinity, higher buffering capacity, and certain therapeutic values in human medicine and nutrition. George Mateljan suggested that doe milk can replace ewe milk or cow milk in diets of those who are allergic to certain mammals' milk. However, like cow milk, doe milk has lactose (sugar), and may cause gastrointestinal problems for individuals with lactose intolerance. In fact, the level of lactose is similar to that of bovine milk.

Fibre

The Angora breed of goats produces long, curling, lustrous locks of mohair. The entire body of the goat is covered with mohair and there are no guard hairs. The locks constantly grow and can be four inches or more in length. Angora crossbreeds, such as the pygora and the nigora, have been created to produce mohair and/or cashgora on a smaller, easier-to-manage animal. The wool is shorn (cut from the body) twice a year, with an average yield of about 10 pounds.

Most goats have softer insulating hairs nearer the skin, and longer guard hairs on the surface. The desirable fibre for the textile industry is the former, and it goes by several names (down, cashmere and pashmina). The coarse guard hairs are of little value as they are too coarse, difficult to spin and difficult to dye. The cashmere goat produces a commercial quantity of cashmere wool, which is one of the most expensive natural fibers commercially produced; cashmere is very fine and soft. The cashmere goat fibre is harvested once a year, yielding around 9 ounces (200 grammes) of down.

In South Asia, cashmere is called "pashmina" (from Persian *pashmina*, "fine wool"). In the 18th and early 19th century, Kashmir (then called Cashmere by the English), had a thriving industry producing shawls from goat down imported from Tibet and Tartary through Ladakh. The shawls were introduced into Western Europe when the General in Chief of the French campaign in Egypt (1799–1802) sent one to Paris. Since these shawls were produced in the upper Kashmir and Ladakh region, the wool came to be known as "cashmere".

Chapter 10

Dairy Cattle Production

Dairy cattle (dairy cows) are cattle cows (adult females) bred for the ability to produce large quantities of milk, from which dairy products are made. Dairy cows generally are of the species *Bos taurus*.

Historically, there was little distinction between dairy cattle and beef cattle, with the same stock often being used for both meat and milk production. Today, the bovine industry is more specialized and most dairy cattle have been bred to produce large volumes of milk. The United States dairy herd produced 83.9 billion kg (185 billion lbs) of milk in 2007, up from 52.6 billion kg (116 billion lbs) in 1950., Yet there are more than 9 million cows on U.S. dairy farms—about 13 million fewer than there were in 1950.

Cow

Dairy cows may be found either in herds on dairy farms where dairy farmers own, manage, care for, and collect milk from them, or on commercial farms. Herd sizes vary around the world depending on landholding culture and social structure. Dairy cow herds in the United States range in size from small farms of a dozen animals to large herds of more than 15,000. The United Kingdom dairy herd overall has nearly 2 million cows, with about 100 head reported on an average farm. In New Zealand, the average herd has more than 375 cows, while in Australia, there are approximately 220 cows in the average herd.

To maintain high milk production, a dairy cow must be bred and produce calves. Depending on market conditions, the cow may be bred with a "dairy bull" or a "beef bull." Female calves (heifers) with dairy breeding may be kept as replacement cows for the dairy herd. If a replacement cow turns out to be a substandard producer of milk, she then goes to market and can be killed for beef. Male calves can either be used later as a breeding bull or sold and used for veal or beef. Dairy farmers usually begin breeding or artificially inseminating heifers around 13 months of age. A cow's gestation period is approximately nine months. Newborn calves are removed from their mothers quickly, usually within three days, as the mother/calf bond intensifies over time and delayed separation can cause extreme stress on the calf.

Domestic cows can live to 20 years, however those raised for dairy rarely live that long, as the average cow is removed from the dairy herd around age four and marketed for beef. In 2009, approximately 19% of the US beef supply came from cull dairy cows: cows that can no longer be seen as an economic asset to the dairy farm. These animals may be sold due to reproductive problems or common diseases of milk cows such as mastitis and lameness.

Cow slaughter is banned in parts of India and remains a contentious issue in states where it is legal. Spent dairy cows don't go to slaughter, but are often seen as roaming on the city streets, and they die of old age or disease. Some pious Hindu organizations manage "old age homes" (Hindi: *Gaushala*) for aged dairy cows.

Calf

Market calves are generally sold at two weeks of age and bull calves may fetch a premium over heifers due to their size, either current or potential. Calves may be sold for veal, or for one of several types of beef production, depending on available local crops and markets. Such bull calves may be castrated if turnout onto pastures is envisaged, in order to render the animals less aggressive. Purebred bulls from elite cows may be put into progeny testing schemes to find out whether they might become superior sires for breeding. Such animals may become extremely valuable.

Dairy CAFO - EPA

Most dairy farms separate calves from their mothers within a day of birth to reduce transmission of disease and simplify management of milking cows. Studies have been done allowing calves to remain with their mothers for 1, 4, 7 or 14 days after birth. Cows whose calves were removed longer than one day after birth showed increased searching, sniffing and vocalizations. However, calves allowed to remain with their mothers for longer periods showed weight gains at three times the rate of early removals as well as more searching behaviour and better social relationships with other calves.

After separation, most young dairy calves subsist on commercial milk replacer, a feed based on dried milk powder. Milk replacer is an economical alternative to feeding whole milk because it is cheaper, can be bought at varying fat and protein percentages. A day old calf consumes around 5 liters of milk per day.

Bull

A bull calf with high genetic potential may be reared for breeding purposes. It may be kept by a dairy farm as a herd bull, to provide natural breeding for the herd cows. A bull may service up to 50 or 60 cows during a breeding season. Any more and the sperm count will decline, leading to cows "returning to service" (to be bred again). A herd bull may only stay for one season since over two years old their temperament becomes too unpredictable. Bull calves intended for breeding commonly are bred on specialized dairy breeding farms, not production farms. These farms are the major source of stocks for artificial insemination.

Milk Production Levels

A cow will produce large amounts of milk over her lifetime. Certain breeds produce more milk than others; however, different breeds produce within a range of around 6,800 to 11,000 kg (15,000 to 25,000 lbs) of milk per lactation. The average for a single dairy cow in the US in 2007 was 9164.4 kg (20,204 lbs) per year, excluding milk consumed by her calves.

Production levels peak at around 40 to 60 days after calving. The cow is then bred. Production declines steadily afterwards, until, at about 305 days after calving, the cow is 'dried off', and milking ceases.

About sixty days later, one year after the birth of her previous calf, a cow will calve again. High production cows are more difficult to breed at a one year interval. Many farms take the view that 13 or even 14 month cycles are more appropriate for this type of cow.

Dairy cows may continue to be economically productive for many lactations. Ten or more lactations are possible. The chances of problems arising which may lead to a cow being culled are high, however; the average herd life of US Holsteins is today fewer than 3 lactations. This requires more herd replacements to be reared or purchased. Over 90% of all cows are culled for 4 main reasons:

Infertility - Failure to Conceive and Reduced Milk Production

Cows are at their most fertile between 60 and 80 days after calving. Cows remaining "open" (not with calf) after this period become increasingly difficult to breed, which may be due to poor health. Failure to expel the afterbirth from a previous pregnancy, luteal cysts, or metritis, an infection of the uterus, are common causes of infertility.

Mastitis - Persistent and Potentially Fatal Mammary Gland Infection, Leading to High Somatic Cell Counts and Loss of Production

Mastitis is recognized by a reddening and swelling of the infected quarter of the udder and the presence of whitish clots or pus in the milk. Treatment is possible with long-acting antibiotics but milk from such cows is not marketable until drug residues have left the cow's system.

Lameness - Persistent Foot Infection or Leg Problems Causing Infertility and Loss of Production

High feed levels of highly digestible carbohydrate cause acidic conditions in the cow's rumen. This leads to laminitis and subsequent lameness, leaving the cow vulnerable to other foot infections and problems which may be exacerbated by standing in feces or water soaked areas.

Production - Some Animals Fail to Produce Economic Levels of Milk to Justify their Feed Costs

Production below 12 to 15 liters of milk per day are not economically viable.

Herd life is strongly correlated with production levels. Lower production cows live longer than high production cows, but may be less profitable. Cows no longer wanted for milk production are sent to slaughter. Their meat is of relatively low value and is generally used for processed meat.

Reproduction

Since the 1950s, artificial insemination (AI) is used at most dairy farms; these farms may keep no bull. Advantages of using AI include its low cost and ease compared to maintaining a bull, ability to select from a large number of bulls to match the anticipated market for the resulting calves, and predictable results.

More recently, embryo transfer has been used to enable the multiplication of progeny from elite cows. Such cows are given hormone treatments to produce multiple embryos. These are then 'flushed' from the cow's uterus. 7-12 embryos are consequently removed from these donor cows and transferred into other cows who serve as surrogate mothers. The result will be between 3 and 6 calves instead of the normal single, or rarely, twins.

Hormone Use

Hormone treatments are given to dairy cows to increase reproduction and to increase milk production.

The hormones are used to produce multiple embryos have to be administered at specific times to dairy cattle to induce ovulation. Frequently, for economic considerations, these drugs are also used to synchronize a group of cows to ovulate simultaneously. The hormones Prostaglandin, Gonadotropin Releasing Hormone, and Progesterone are used for this purpose and sold under the brand names Lutalyse, Cystorelin, Estrumate, Factrel, Prostamate, Fertagyl. Insynch, and Ovacyst. They may be administered by injection, insertion or mixed with feed.

About 17% of dairy cows in the United States are injected with Bovine somatotropin, also called recombinant bovine somatotropin (rBST), recombinant bovine growth hormone (rBGH), or artificial growth hormone. The use of this hormone increases milk production from 11%-25%, but also increases the likelihood of cattle developing mastitis, reduction in fertility and lameness. The U.S. Food and Drug Administration (FDA) has ruled that rBST is harmless to people, although critics point out increased levels of insulin-like growth factor 1 (IGF-1) in milk produced using this hormone. The use of rBST is banned in Canada, parts of the European Union, Australia and New Zealand.

Nutrition

Nutrition plays an important role in keeping cattle healthy and strong. Implementing an adequate nutrition program can also improve milk production and reproductive performance. Nutrient requirements may not be the same depending on the animal's age and stage of production.

Forages, which refer especially to hay or straw, are the most common type of feed used. Cereal grains, as the main contributors of starch to diets, are important in meeting the energy needs of dairy cattle. Barley is one example of grain that is extensively used around the world. Barley is grown in temperate to sub-artic climates, and it is transported to those areas lacking the necessary amounts of grain. Although variations may occur, in general, barley is an excellent source of balanced amounts of protein, energy, and fibre.

Ensuring adequate body fat reserves is essential for cattle to produce milk and also to keep reproductive efficiency. However, if cattle get excessively fat or too thin, they run the risk of developing metabolic problems.

Scientists have found that a variety of fat supplements can benefit conception rates of lactating dairy cows. Some of these different fats include oleic acids, found in canola oil, animal tallow, and yellow grease; palmitic acid found in granular fats and dry fats; and linolenic acids which are found in cottonseed, safflower, sunflower, and soybean. It is also important to note that proper levels of fat also improve cattle longevity.

Using by-products is one way of reducing the normally high feed costs. However, lack of knowledge of their nutritional and economic value limits their use. Although the reduction of costs may be significant, they have to be used carefully because animal may have negative reactions to radical changes in feeds, (e.g. fog fever). Such a change must then be made slowly and with the proper follow up.

Pesticide Use

A survey of the primary dairy producing areas in the US indicated that 13 percent of lactating animals were treated with insecticides permethrin, pyrethrin, coumaphos, and dichlorvos primarily by daily or every-other-day coat sprays. Workers, particularly in stanchion barns, may be exposed to higher than recommended amounts of these pesticides.

Breeds

In the United States, dairy cattle are divided into six major breeds. These are the: Holstein-Friesian, Brown Swiss, Guernsey, Ayrshire, Jersey, and Milking Shorthorn.

In Rajasthan, an indigenous breed called *Tharparkar* exists, named from the Tharparkar District, now in Sindh Pakistan. Another type of dairy cow known as *Nagauri* from Nagaur District, the bull of which is renowned for its ability to plow fields and run. Traditionally, they used to pull covered wagons, known as *rath*, and in marriages to transport the newlywed couple. They are now a crutch for thriving agricultural and livestock rearing societies of the Thar Desert.

Many other breeds are used nearly exclusively for beef, or for both dairy and beef purposes.

Artificial Insemination

Artificial insemination, or AI, is the process by which sperm is placed into the reproductive tract of a female for the purpose of impregnating the female by using means other than sexual intercourse or natural insemination. In humans, it is used as assisted reproductive technology, using either sperm from the woman's male partner or sperm from a sperm donor (donor sperm) in cases where the male partner produces no sperm or the woman has no male partner (i.e., single women, lesbians). In cases where donor sperm is used the woman is the gestational and genetic mother of the child produced, and the sperm donor is the genetic or biological father of the child.

Artificial insemination is widely used for livestock breeding, especially for dairy cattle and pigs. Techniques developed for livestock have been adapted for use in humans.

Specifically, freshly ejaculated sperm, or sperm which has been frozen and thawed, is placed in the cervix (intracervical insemination – ICI) or, after washing, into the female's uterus (intrauterine insemination – IUI) by artificial means.

In humans, artificial insemination was originally developed as a means of helping couples to conceive where there were 'male factor' problems of a physical or psychological nature affecting the male partner which prevented or impeded conception. Today, the process is also and more commonly used in the case of choice mothers, where a woman has no male partner and the sperm is provided by a sperm donor.

In Humans

Preparations

A sperm sample will be provided by the male partner of the woman undergoing artificial insemination, but sperm provided through sperm donation by a sperm donor may be used if, for example, the woman's partner produces too few motile sperm, or if he carries a genetic disorder, or if the woman has no male partner. Sperm is usually obtained through masturbation or the use of an electrical stimulator, although a special condom, known as a collection condom, may be used to collect the semen during intercourse.

The man providing the sperm is usually advised not to ejaculate for two to three days before providing the sample in order to increase the sperm count.

A woman's menstrual cycle is closely observed, by tracking basal body temperature (BBT) and changes in vaginal mucus, or using ovulation kits, ultrasounds or blood tests.

When using intrauterine insemination (IUI), the sperm must have been "washed" in a laboratory and concentrated in Hams F10 media without L-glutamine, warmed to 37C. The process of "washing" the sperm increases the chances of fertilization and removes any mucus and non-motile sperm in the semen. Pre and post concentration of motile sperm is counted.

If sperm is provided by a sperm donor through a sperm bank, it will be frozen and quarantined for a particular period and the donor will be tested before and after production of the sample to ensure that he does not carry a transmissible disease. Sperm samples donated in this way are produced through masturbation by the sperm donor at the sperm bank. A chemical known as a cryoprotectant is added to the sperm to aid the freezing and thawing process. Further chemicals may be added which separate the most active sperm in the sample as well as extending or diluting the sample so that vials for a number of inseminations are produced. For fresh shipping, a semen extender is used.

If sperm is provided by a private donor, either directly or through a sperm agency, it is usually supplied fresh, not frozen, and it will not be quarantined. Donor sperm provided in this way may be given directly to the recipient woman or her partner, or it may be transported in specially insulated containers. Some donors have their own freezing apparatus to freeze and store their sperm. Private donor sperm is usually produced through masturbation, but some donors use a collection condom to obtain the sperm when having sexual intercourse with their own partners.

Procedure

When an ovum is released, semen provided by the woman's male partner, or by a sperm donor, is inserted into the woman's vagina or uterus. The semen may be fresh or it may be frozen semen which has been thawed. Where donor sperm is supplied by a sperm bank, it will always be quarantined and frozen and will need to be thawed before use. Specially designed equipment is available for carrying out artificial inseminations. In the case of vaginal artificial insemination, semen is usually placed in the vagina by way of a needleless syringe. A longer tube, known as a 'tom cat' may be attached to the end of the syringe to facilitate deposit of the semen deeper into the vagina. The woman is generally advised to lie still for a half hour or so after the insemination to prevent seepage and to allow fertilization to take place.

A more efficient method of artificial insemination is to insert semen directly into the woman's uterus. Where this method is employed it is important that only 'washed' semen be used and this is inserted into the uterus by means of a catheter. Sperm banks and fertility clinics usually offer 'washed' semen for this purpose, but if partner sperm is used it must also be 'washed' by a medical practitioner to eliminate the risk of cramping.

Semen is occasionally inserted twice within a 'treatment cycle'. A double intrauterine insemination has been theorized to increase pregnancy rates by decreasing the risk of missing the fertile window during ovulation. However, a randomized trial of insemination after ovarian hyperstimulation found no difference in live birth rate between single and double intrauterine insemination.

An alternative method to the use of a needless syringe or a catheter involves the placing of partner or donor sperm in the woman's vagina by means of a specially designed cervical cap, a conception device or conception cap. This holds the semen in place near to the entrance to the cervix for a period of time, usually for several hours, to allow fertilization to take place. Using this method, a woman may go about her usual activities while the cervical cap holds the semen in the vagina. One advantage with the conception device is that fresh, non-liquified semen may be used.

If the procedure is successful, the woman will conceive and carry to term a baby. A pregnancy resulting from artificial insemination will be no different from a pregnancy achieved by sexual intercourse. However, there may be a slight increased likelihood of multiple births if drugs are used by the woman for a 'stimulated' cycle.

Donor Variations

Either sperm provided by the woman's husband or partner (artificial insemination by husband, AIH) or sperm provided by a known or anonymous sperm donor (artificial insemination by donor, AID or DI) can be used.

Techniques

Intrauterine insemination, Intravaginal insemination, Intracervical insemination, and Intratubal insemination.

Intracervical Insemination

ICI is the easiest way to inseminate. This involves the deposit of raw fresh or frozen semen (which has been thawed) by injecting it high into the cervix with a needle-less syringe. This process closely replicates the way in which fresh semen is directly deposited on to the neck of the cervix by the penis during vaginal intercourse. When the male ejaculates, sperm deposited this way will quickly swim into the cervix and toward the fallopian tubes where an ovum recently released by the ovary(s) hopefully awaits fertilization. It is the simplest method of artificial insemination and 'unwashed' or raw semen is normally used. It is probably therefore, the most popular method and is used in most home, self and practitioner insemination procedures.

Timing is critical as the window and opportunity for fertilization, is little more than 12 hours from the release of the ovum. For each woman who goes through this process be it AI (artificial insemination) or NI (natural insemination); to increase chances for success, an understanding of her rhythm or natural cycle is very important. Home ovulation tests are now available. Doing and understanding Basal Temperature Tests over several cycles; there is a slight dip and quick rise at the time of ovulation. She should note the color and texture of her vaginal mucous discharge. At the time of ovulation the protective cervical plug is released giving the vaginal discharge a stringy texture with an egg white color. A woman may also be able check the softness of the nose of her cervix by inserting two fingers. It should be considerably softer and more pliable than normal.

Advanced technical (medical) procedures may be used to increase the chances of conception.

When performed at home without the presence of a professional this procedure is sometimes referred to as intravaginal insemination or IVI.

Intrauterine Insemination

'Washed sperm', that is, spermatozoa which have been removed from most other components of the seminal fluids, can be injected directly into a woman's uterus in a process called intrauterine insemination (IUI). If the semen is not washed it may elicit uterine cramping, expelling the semen and causing pain, due to content of prostaglandins. (Prostaglandins are also the compounds responsible for causing the myometrium to contract and expel the menses from the uterus, during menstruation.) The woman should rest on the table for 15 minutes after an IUI to optimize the pregnancy rate.

To have optimal chances with IUI, the female should be under 30 years of age, and the man should have a TMS of more than 5 million per ml. In practice, donor sperm will satisfy these criteria. A promising cycle is one that offers two follicles measuring more than 16 mm, and estrogen of more than 500 pg/mL on the day of hCG administration. A short period of ejaculatory abstinence before intrauterine insemination is associated with higher pregnancy rates. However, GnRH agonist administration at the time of implantation does not improve pregnancy outcome in intrauterine insemination cycles according to a randomized controlled trial.

It can be used in conjunction with ovarian hyperstimulation. Still, advanced maternal age causes decreased success rates; Women aged 38–39 years appear to have reasonable success during the first two cycles of ovarian hyperstimulation and IUI. However, for women aged e"40 years, there appears to be no benefit after a single cycle of COH/IUI. It is therefore recommended to consider in vitro fertilization after one failed COH/IUI cycle for women aged e"40 years.

Intrauterine Tuboperitoneal Insemination

Intrauterine tuboperitoneal insemination (IUTPI) is insemination where both the uterus and fallopian tubes are filled with insemination fluid. The cervix is clamped to prevent leakage to the vagina, best achieved with the specially designed Double Nut Bivalve (DNB) speculum. The sperm is mixed to create a volume of 10 ml, sufficient enough to fill the uterine cavity, pass through the interstitial part of the tubes and the ampulla, finally reaching the peritoneal cavity and the Pouch of Douglas where it would be mixed with the peritoneal and follicular fluid. IUTPI can be useful in unexplained infertility, mild or moderate male infertility, and mild or moderate endometriosis.

Intratubal Insemination

IUI can furthermore be combined with intratubal insemination (ITI), into the Fallopian tube although this procedure is no longer generally regarded as having any beneficial effect compared with IUI. ITI however, should not be confused with gamete intrafallopian transfer, where both eggs and sperm are mixed outside the woman's body and then immediately inserted into the Fallopian tube where fertilization takes place.

Pregnancy Rate

Success rates, or pregnancy rates for artificial insemination may be very misleading, since many factors including the age and health of the recipient have to be included to give a meaningful answer, e.g. definition of success and calculation of the total population. For couples with unexplained infertility, unstimulated IUI is no more effective than natural means of conception.

Approximate pregnancy rate as a function of total sperm count (may be twice as large as total motile sperm count). Values are for intrauterine insemination. (Old data, rates are likely higher today)

Generally, it is 10 to 15% per menstrual cycle using ICI, and and 15-20% per cycle for IUI. In IUI, about 60 to 70% have achieved pregnancy after 6 cycles.

As seen on the graph, the pregnancy rate also depends on the total sperm count, or, more specifically, the total motile sperm count (TMSC), used in a cycle. It increases with increasing TMSC, but only up to a certain count, when other factors become limiting to success. The summed pregnancy rate of two cycles using a TMSC of 5 million (may be a TSC of ~10 million on graph) in each cycle is substantially higher than one single cycle using a TMSC of 10 million. However, although more cost-efficient, using a lower TMSC also increases the average time taken before getting pregnant. Women whose age is becoming a major factor in fertility may not want to spend that extra time.

Samples Per Child

How many samples (ejaculates) that are required give rise to a child varies substantially from person to person, as well as from clinic to clinic. However, the following equations generalize the main factors involved: For intracervical insemination:

$$N = \frac{V_s \times c \times r_s}{n_r}$$

- N is how many children a single sample can give rise to.
- V_s is the volume of a sample (ejaculate), usually between 1.0 mL and 6.5 mL
- c is the concentration of motile sperm in a sample *after freezing and thawing*, approximately 5-20 million per ml but varies substantially
- r_s is the pregnancy rate per cycle, between 10% to 35%

- n_r is the total motile sperm count recommended for vaginal insemination (VI) or intra-cervical insemination (ICI), approximately 20 million pr. ml.

The pregnancy rate increases with increasing number of motile sperm used, but only up to a certain degree, when other factors become limiting instead.

In the simplest form, the equation reads:

$$N = \frac{n_s}{n_c} \times r_s$$

N is how many children a single sample can give rise to

n_s is the number of vials produced per sample

n_c is the number of vials used in a cycle

r_s is the pregnancy rate per cycle

n_s can be further split into:

$$n_s = \frac{V_s}{V_v}$$

n_s is the number of vials produced per sample

V_s is the volume of a sample

V_v is the volume of the vials used

n_c may be split into:

$$n_c = \frac{n_r}{n_s}$$

n_c is the number of vials used in a cycle

n_r is the number of motile sperm recommended for use in a cycle

n_s is the number of motile sperm in a vial

n_s may be split into:

$$n_s = V_v \times c$$

n_s is the number of motile sperm in a vial

V_v is the volume of the vials used

c is the concentration of motile sperm in a sample

Thus, the factors can be presented as follows:

$$N = \frac{V_s \times c \times r_s}{n_r} \times \frac{V_v}{V_v}$$

N is how many children a single sample can help giving rise to

V_s is the volume of a sample

c is the concentration of motile sperm in a sample

r_s is the pregnancy rate per cycle

n_r is the number of motile sperm recommended for use in a cycle

V_v is the volume of the vials used (its value doesn't affect *N* and may be eliminated. In short, the smaller the vials, the more vials are used)

Approximate live birth rate (r_s) among infertile couples as a function of total motile sperm count (n_r). Values are for intrauterine insemination.

With these numbers, one sample would on average help giving rise to 0.1-0.6 children, that is, it actually takes on average 2-5 samples to make a child.

For intrauterine insemination (IUI), a *centrifugation fraction* (f_c) may be added to the equation:

f_c is the fraction of the volume that remains after centrifugation of the sample, which may be about half (0.5) to a third (0.33).

$$N = \frac{V_s \times f_c \times c \times r_s}{n_r}$$

On the other hand, only 5 million motile sperm may be needed per cycle with IUI (n_r=5 million)

Thus, only 1-3 samples may be needed for a child if used for IUI.

History

In the 1970s, direct intraperitoneal insemination (DIPI) was occasionally used, where doctors injected sperm into the lower abdomen through a surgical hole or incision, with the intention of letting them find the oocyte at the ovary or after entering the genital tract through the ostium of the fallopian tube.

Artificial Insemination in Livestock and Pets

Artificial insemination is used in many non-human animals, including sheep, horses, cattle, pigs, dogs, pedigree animals generally, zoo animals, turkeys and even honeybees. It may be used for many reasons, including to allow a male to inseminate a much larger number of females, to allow use of genetic material from males separated by distance or time, to overcome physical breeding difficulties, to control the paternity of offspring, to synchronise births, to avoid injury incurred during natural mating, and to avoid the need to keep a male at all (such as for small numbers of females or in species whose fertile males may be difficult to manage).

Semen is collected, extended, then cooled or frozen. It can be used on site or shipped to the female's location. If frozen, the small plastic tube holding the semen is referred to as a *straw*. To allow the sperm to remain viable during the time before and after it is frozen, the semen is mixed with a solution containing glycerol or other cryoprotectants. An *extender* is a solution that allows the semen from a donor to impregnate more females by making insemination possible with fewer sperm. Antibiotics, such as streptomycin, are sometimes added to the sperm to control some bacterial venereal diseases. Before the actual insemination, estrus may be induced through the use of progestogen and another hormone (usually PMSG).

Artificial insemination of farm animals is very common in today's agriculture industry in the developed world, especially for breeding dairy cattle. (75% of all inseminations) Swine are also bred using this method (up to 85% of all inseminations). It provides an economical means for a livestock breeder to improve their herds utilizing males having very desirable traits.

Although common with cattle and swine, AI is not as widely practised in the breeding of horses. A small number of equine associations in North America accept only horses that have been conceived by "natural cover" or "natural service" – the actual physical mating of a mare to a stallion. The Jockey Club being the most notable of these - no AI is allowed in Thoroughbred breeding. Other registries such as the AQHA and warmblood registries allow registration of foals created through AI, and the process is widely used allowing the breeding of mares to stallions not resident at the same facility - or even in the same country - through the use of transported frozen or cooled semen.

Modern Artificial Insemination was pioneered by Dr. John O. Almquist of the Pennsylvania State University. His improvement of breeding efficiency by the use of antibiotics (first proven with penicillin in 1946) to control bacterial growth, decreasing embrionic mortality and increase fertiilty, and various new techniques for processing, freezing and thawing of frozen semen significantly enhanced the practical utilization of AI in the livestock industry, and earned him the 1981 Wolf Foundation Prize in Agriculture. Many techniques developed by him have since been applied to other species, including that of the human male.

The, John O. Almquist Dairy Breeding Research Center, at Penn State University, is a unique facility for research on reproduction in farm animals. It is one of only three centers worldwide providing substantial housing for mature bulls. The Center has a long-standing working relationship with the artificial insemination industry.

Embryo Transfer

Embryo transfer refers to a step in the process of assisted reproduction in which one or several embryos are placed into the uterus of a female with the intent to establish a pregnancy. This technique (which is often used in connection with in vitro fertilization (IVF)), may be used in humans or in animals, in which situations the goals may vary.

Fresh Versus Frozen

Embryos can be either "fresh" from fertilized egg cells of the same menstrual cycle, or "frozen", that is they have been generated in a preceding cycle, cryopreserved, and are thawed just prior to the transfer. Babies born from frozen IVF embryos are less likely to be born prematurely or underweight than are those conceived during fresh treatment cycles, three independent teams of scientists have found. One of the studies also recorded lower rates of stillbirth and early death among frozen-embryo babies. The results, from researchers based in the United States, Australia and Finland, suggest that far from being riskier than conventional IVF, as is generally thought, cycles using frozen embryos may actually be safer, Mark Henderson of the Times London reported in November 2008.

Uterine Preparation

In the human, the uterine lining (endometrium) needs to be appropriately prepared so that the embryo(s) can implant. In a natural or stimulated cycle, the embryo transfer takes place in the luteal phase at a time where the lining is appropriately undeveloped in relation to the status of the present Luteinizing Hormone. In a cycle where a "frozen" embryo is transferred, the recipient woman could be given first estrogen preparations (about 2 weeks), then a combination of oestrogen and progesterone so that the lining becomes receptive for the embryo. The time of receptivity is the implantation window.

Timing

In stimulated cycles in human IVF, embryos are typically transferred 3 days after fertilization and may then be at the eight-cell stage, or they are transferred 2 to 3 days later when they have reached the blastocyst stage. Embryos who reach the day 3 cell stage can be tested for chromosal or specific genetic defects prior to possible transfer by preimplantation genetic diagnosis (PGD).

Monozygotic twinning is not increased after blastocyst transfer compared with cleavage-stage embryo transfer.

Procedure

The embryo transfer procedure starts by placing a speculum in the vagina to visualize the cervix, which is cleansed with saline solution or culture media. A soft transfer catheter is loaded with the embryos and handed to the clinician after confirmation of the patient's identity. The catheter is inserted through the cervical canal and advanced into the uterine cavity.

There is good and consistent evidence of benefit in *ultrasound guidance*, that is, making an abdominal ultrasound to ensure correct placement, which is 1–2 cm from the uterine fundus. Anesthesia is generally not required. Single embryo transfers in particular require accuracy and precision in placement within the uterine cavity. The optimal target for embryo placement, known as the maximal implantation potential (MIP) point, is identified using 3D/4D ultrasound. However, there is limited evidence that supports deposition of embryos in the midportion of the uterus.

After insertion of the catheter, the contents are expelled and the embryos are deposited. Limited evidence supports making trial transfers before performing the procedure with embryos. After expulsion, the duration that the catheter remains inside the uterus has no effect on pregnancy rates. Limited evidence suggests avoiding negative pressure from the catheter after expulsion. After withdrawal, the catheter is handed to the embryologist, who inspects it for retained embryos. In the process of zygote intrafallopian transfer (ZIFT), eggs are removed from the woman, fertilised, and then placed in the woman's fallopian tubes rather than the uterus.

Embryo Number

A major issue is how many embryos should be transferred. Placement of multiple embryos carries the risk of multiple pregnancy.

In the past, physicians have often placed too many embryos in the hope to establish a pregnancy. However, the rise in multiple pregnancies has led to a reassessment of this approach. Professional societies and in many countries, the legislature, have issued guidelines or laws to curtail a practice of placing too many embryos in an attempt to reduce multiple pregnancies.

E-SET

The technique of selecting only one embryo to transfer to the woman is called elective-Single Embryo Transfer (e-SET) or, when embryos are at the blastocyst stage, it can also be called *elective single blastocyst transfer (eSBT).* It lowers the risk of multiple pregnancies, compared with e.g. Double Embryo Transfer (DET) or *double blastocyst transfer* (2BT), with a twinning rate of approximately 3.5% in sET compared with approximately 38% in DET, or 2% in eSBT compared with approximately 25% in 2BT. At the same time, pregnancy rates is not significantly less with eSBT than with 2BT. Furthermore, SET has better outcomes in terms of mean gestational age at delivery, mode of delivery, birthweight, and risk of neonatal intensive care unit necessity than DET. e-SET of embryos at the cleavage stage reduces the likelihood of live birth by 38% and multiple birth by 94%. Evidence from randomized, controlled trials suggests that increasing the number of e-SET attempts (fresh and/or frozen) results in a cumulative live birth rate similar to that of DET.

The usage of single embryo transfer is highest in Sweden (69.4%), but as low as 2.8% in the USA. Access to public funding for ART, availability of good cryopreservation facilities and legislation appear to be the most important factors for regional usage of single embryo transfer. Also, personal choice plays a significant role as many subfertile couples have a strong preference for twins.

Follow-up

Patients usually start progesterone medication after egg (also called oocyte) retrieval. While daily intramuscular injections of progesterone-in-oil (PIO) have been the standard route of administration, PIO injections are not FDA-approved for use in pregnancy. A recent meta-analysis showed that the intravaginal route with an appropriate dose and dosing frequency is equivalent to daily intramuscular injections. In addition, a recent case-matched study comparing vaginal progesterone with PIO injections showed that live birth rates were nearly identical with both methods. A duration of progesterone administration of 11 days results in almost the same birth rates as longer durations. Patients are also given estrogen medication in some cases after the embryo transfer. Pregnancy testing is done typically two weeks after egg retrieval.

Third-party Reproduction

It is not necessary that the embryo transfer be performed on the female who provided the eggs. Thus another female whose uterus is appropriately prepared can receive the embryo and become pregnant. Embryo transfer may be used where a woman who has eggs but no uterus and wants to have a biological baby; she would require the help of a gestational carrier or surrogate to carry the pregnancy. Also, a woman who has no eggs but a uterus may resort to egg donor IVF, in which case another woman would provide eggs for fertilization and the resulting embryos are placed into the uterus of the patient. Fertilization may be performed using the woman's partner's sperm or by using donor sperm. 'Spare' embryos which are created for another couple undergoing IVF treatment but which are then surplus to that couple's needs may also be transferred.

Embryos may be specifically created by using eggs and sperm from donors and these can then be transferred into the uterus of another woman. A surrogate may carry a baby produced by embryo transfer for another couple, even though neither she nor the 'commissioning' couple is biologically related to the child. Third party reproduction is controversial and regulated in many countries. Persons entering gestational surrogacy arrangements must make sense of an entirely new type of relationship that does not fit any of the traditional scripts we use to categorize relations as kinship, friendship, romantic partnership or market relations. Surrogates have the experience of carrying a baby that they conceptualize as not of their own kin, while intended mothers have the experience of waiting through nine months of pregnancy and transitioning to motherhood from outside of the pregnant body. This can lead to new conceptualizations of body and self.

History

The first transfer of an embryo from one human to another resulting in pregnancy was reported in July 1983 and subsequently led to the announcement of the first human birth February 3, 1984. This procedure was performed at the Harbor UCLA Medical Center under the direction of Dr. John Buster and the University of California at Los Angeles School of Medicine.

In the procedure, an embryo that was just beginning to develop was transferred from one woman in whom it had been conceived by artificial insemination to another woman who gave birth to the infant 38 weeks later. The sperm used in the artificial insemination came from the husband of the woman who bore the baby.

This scientific breakthrough established standards and became an agent of change for women suffering from the afflictions of infertility and for women who did not want to pass on genetic disorders to their children. Donor embryo transfer has given women a mechanism to become pregnant and give birth to a child that will contain their husband's genetic makeup. Although donor embryo transfer as practiced today has evolved from the original non-surgical method, it now accounts for approximately 5% of in vitro fertilization recorded births.

Prior to this, thousands of women who were infertile, had adoption as the only path to parenthood. This set the stage to allow open and candid discussion of embryo donation and transfer. This breakthrough has given way to the donation of human embryos as a common practice similar to other donations such as blood and major organ donations. At the time of this announcement the event was captured by major news carriers and fueled healthy debate and discussion on this practice which impacted the future of reproductive medicine by creating a platform for further advancements in woman's health.

This work established the technical foundation and legal-ethical framework surrounding the clinical use of human oocyte and embryo donation, a mainstream clinical practice, which has evolved over the past 25 years. Building upon this groundbreaking research and since the initial birth announcement in 1984, well over 47,000 live births resulting from donor embryo transfer have been and continue to be recorded by the Centers for Disease Control(CDC) in the United States to infertile women, who otherwise would not have had children by any other existing method.

Embryo Transfer in Livestock

Embryo transfer techniques allow top quality female livestock to have a greater influence on the genetic advancement of a herd or flock in much the same way that artificial insemination has allowed greater use of superior sires. ET also allows the continued use of animals such as competition mares to continue training and showing, while producing foals. The general epidemiological aspects of embryo transfer indicates that the transfer of embryos provides the opportunity to introduce genetic material into populations of livestock while greatly reducing the risk for transmission of infectious diseases. Recent developments in the sexing of embryos before transfer and implanting has great potential in the dairy and other livestock industries.

Products and Adaptations of Dairy Cattle

Transition cow biology and management has become a focal point for research in nutrition and physiology during the past 15 yr. First, it was recognized that many of the metabolic disorders afflicting cows during the periparturient period are interrelated in their occurrence and are related to the diet fed during the prepartum period (Curtis et al., 1985).

They determined that increased energy content of the diet fed during the prepartum period was associated with decreased incidence of displaced abomasum and that increased protein content of this diet was associated with decreased incidences of retained placenta and ketosis (Curtis et al., 1985). Although the strategy for prevention of milk fever was to feed a prepartum diet low in Ca at that time, Ca content of the prepartum diet was not related to the occurrence of milk fever in their study. These results led to substantial investigation of the biological relationships underpinning these epidemiological relationships.

Despite the prodigious output of research on the nutrition and physiology of transition cows, the transition period remains a problematic area on many dairy farms, and metabolic disorders continue to occur at economically important rates on commercial dairy farms (Burhans et al., 2003). Data recently summarized (Godden et al., 2003) indicate that approximately 25% of cows that left dairy herds in Minnesota from 1996 to 2001 did so during the first 60 DIM, with an uncertain additional percentage leaving by the end of the lactation due in part to difficulty during the transition period. The economic ramifications of the loss of cows early in lactation together with the comprehensive costs associated with occurrence of the various metabolic disorders in both clinical and subclinical form are large. Therefore, research attention will continue to focus on understanding the biology of transition cows and implementing management schemes on dairy farms to optimize production and profitability on these farms. Excellent reviews have been written to describe the adaptations in energy, protein, and mineral metabolism that must occur for dairy cows to transition successfully to lactation and the authors of NRC (2001) effectively integrated the metabolic adaptations described by these authors with nutritional recommendations. Hence, the purpose of this review is to briefly overview our knowledge of these metabolic adaptations and then to update knowledge summarized by the NRC (2001) with more recent research on transition cow nutrition and management. Furthermore, recent research integrating immune function with metabolism in dairy cows will be overviewed, with emphasis on the implications of this interrelationship for the successful implementation of programs for transition cows on dairy farms. Given our emphasis on incorporating the most recent information, we have chosen to include abstracted results to extend the peer-reviewed literature where appropriate. We acknowledge that this approach carries some risk and limitations, but believe that it is consistent with our intent to provide as current a view of this rapidly evolving area as possible.

Metabolic Adaptations During

The Transition Period

The hallmark of the transition period of dairy cattle is the dramatic change in nutrient demands that necessitate exquisite coordination of metabolism to meet requirements for energy, glucose, AA, and Ca by the mammary gland following calving. Estimates of the demand for glucose, AA, fatty acids, and net energy by the gravid uterus at 250 d of gestation and the lactating mammary gland at 4 d postpartum indicate approximately a tripling of demand for glucose, a doubling of demand for AA, and approximately a fivefold increase in demand for fatty acids during this timeframe (Bell, 1995). In addition, the requirement for Ca increases approximately fourfold on the day of parturition (Horst et al., 1997). The cow relies on homeorhetic controls to enable these changes in nutrient partitioning to occur.

Glucose Metabolism

The primary homeorhetic adaptation of glucose metabolism to lactation is the concurrent increase in hepatic gluconeogenesis (Reynolds et al., 2003) and decrease in oxidation of glucose by peripheral tissues (Bennink et al., 1972) to direct glucose to the mammary gland for lactose synthesis. Reynolds et al. (2003) reported that net flux of glucose across the portal-drained viscera of cows was zero to slightly negative during the transition period and early lactation; the 267% increase in total splanchnic output of glucose from 9 d before expected parturition to 21 d after parturition resulted almost completely from increased hepatic gluconeogenesis. The major substrates for hepatic gluconeogenesis in ruminants are propionate from ruminal fermentation, lactate from Cori cycling, AA from protein catabolism or net portaldrained visceral absorption, and glycerol released during lipolysis in adipose tissue (Seal and Reynolds, 1993). The maximal calculated contribution of propionate to net glucose release by liver ranged from approximately 50 to 60% during the transition period; that for lactate ranged from 15 to 20%; and that for glycerol ranged from 2 to 4% (Reynolds et al., 2003). By difference, AA accounted for a minimum of approximately 20 to 30% during the transition period; the maximal contribution of Ala increased from 2.3% at 9 d prepartum to 5.5% at 11 d postpartum. These results are consistent with those of Overton et al. (1998), who reported that hepatic capacity to convert [1-14C]alanine to glucose was approximately doubled on 1 d postpartum compared with 21 d prepartum. Although AA are not likely to be quantitatively important in terms of the amount of milk that the AA pool will support during early lactation, these results lend support to the use of AA as an adaptational substrate pool for glucose synthesis during the immediate postpartum period.

Lipid Metabolism

The primary homeorhetic adaptation of lipid metabolism to lactation is the mobilization of body fat stores to meet the overall energetic requirements of the cow during a period of negative energy balance in early lactation. Body fat is mobilized into the bloodstream in the form of NEFA. The NEFA are utilized to make upwards of 40% of milk fat during the first days of lactation (Bell, 1995). Skeletal muscle uses someNEFAfor fuel, particularly as it decreases its reliance on glucose as a fuel during early lactation. Given that plasma NEFA concentrations increase in response to increased energy needs accompanied by inadequate feed intake, DMI and plasma NEFA concentrations usually are inversely related.

Available evidence suggests that the liver takes up NEFA in proportion to their supply, but the liver typically does not have sufficient capacity to completely dispose of NEFA through export into the blood or catabolism for energy; therefore, cows are predisposed to accumulate NEFA as triglycerides within liver when large amounts of NEFA are released from adipose tissue into the circulation (Emery et al., 1992). It is likely that some triglyceride accumulates in liver of almost all high-producing cows during the first few weeks postpartum. What is uncertain is the threshold at which fat begins to have detrimental effects on other hepatic processes.

Piepenbrink and Overton (2003b) reported that there is a negative correlation (r = “0.4) between triglyceride accumulation in liver and the capacity of liver slices to convert propionate to glucose in vitro. Cadorniga-Valino et al. (1997) demonstrated that lipid infiltration of isolated hepatocytes decreased gluconeogenic capacity from propionate. A subsequent experiment using a "physiological" mixture of fatty acids determined that lipid infiltration did not affect rates of gluconeogenesis but decreased ureagenic capacity (Strang et al., 1998). The implications of decreased ureagenic capacity are not clear, but limited evidence suggests that this phenomenon may occur in dairy cows during the transition period. Zhu et al. (2000) determined that peripheral concentrations of ammonia doubled when liver triglyceride concentrations increased during the first 2 d postpartum. In vitro incubation with ammonium chloride strongly inhibited capacity of isolated hepatocytes to synthesize glucose from propionate (Overton et al., 1999). Therefore, it is conceivable that inhibition of gluconeogenesis may occur in vivo when triglycerides accumulate in liver. Perhaps the mechanism is modulated by ammonia supply to liver. Potential implications of this research for the management of transition dairy cows centers around carbohydrate and protein nutrition. As discussed previously, significant quantities of AA are needed for gluconeogenesis. However, we hypothesize that excess protein or asynchronous supply of ruminalNrelative to carbohydrate supply may increase the ammonia load on the animal, and thereby affect the capacity of a triglycerideladen liver to synthesize glucose. Furthermore, the specific nature of the potential link between impaired gluconeogenic capacity and ureagenic capacity has yet to be elucidated.

Calcium Metabolism

Reinhardt et al. (1988) reviewed basic Ca, P, and Mg homeostatic mechanisms in ruminants, and further re view of Ca and vitamin D metabolism in dairy cows were given by Horst et al. (1994). The skeleton contains 99 and 80% of total body Ca and P, respectively. Calcium pools are under strict homeostatic control, whereas the P pool is less regulated. Under noninflammatory physiologic conditions, serum Ca and P concentrations are under endocrine control regulated at the level of intestinal absorption, bone resorption or deposition, renal reabsorption and urinary excretion, salivary recycling, fetal deposition (pregnant animal), milk secretion (lactating animal), and fecal excretion.

In the absence of inflammation, parathyroid hormone and 1,25-dihydroxyvitamin D are generally conservatory for the extracellular pool and are responsible for increasing intestinal absorption and renal reabsorption of Ca, and bone resorption of Ca and P. Although parathyroid hormone adds to the extracellular phosphate pool via bone resorption, it actually increases renal phosphate excretion and, more significantly, increases salivary phosphate secretion. Parathyroid hormone-related protein may also be important for the secretion of Ca (as well as Mg and P) into the milk of lactating animals (Thiede, 1994). Calcitonin is secreted by the thyroid gland in response to elevated serum Ca and results in increased bone mineral deposition, decreased intestinal absorption, and increased urinary Ca excretion. Study of mastectomized cows (Goff et al., 2002) indicated that the mammary gland and its concomitant lactogenesis are fully responsible for the periparturient hypocalcemia. In contrast, serum P concentration decreased irrespective of mastectomy, indicating that factors other than milk production at the time of parturition are responsible for periparturient hypophosphatemia. Although circulating concentrations of regulatory hormones yield some information about macromineral homeostasis, these data alone may be insufficient to elucidate the mechanisms of macromineral dysregulation. For example, plasma concentrations of parathyroid hormone (Mayer et al., 1969) and 1,25-dihydroxyvitamin D (Horst et al., 1978) are actually elevated, while plasma calcitonin concentration is decreased (Mayer et al., 1975) immediately preceding, and during, most cases of hypocalcemic parturient paresis in dairy cows.

Thus, factors at the tissue level other than hormone concentrations, such as receptor numbers, binding affinity, hormone clearance, and postreceptor signaling, may also be affected during cases of macromineral dysregulation (Horst et al., 1994). Nutritional strategies to minimize periparturient hypocalcemia are based on manipulation of these endocrine control-points by priming the absorptive and resorptive mechanisms of macromineral metabolism so that the cow can more efficiently manage the period of negative mineral balance associated with the onset of lactation (Horst et al., 1997).

Nutritional Management to Support Metabolic Adaptations During the Transition Period

Grouping Strategies

The primary goal of nutritional management strategies of dairy cows during the transition period should be to support the metabolic adaptations described above. Industry-standard nutritional management of dairy cows during the dry period consists of a 2-group nutritional scheme. The NRC (2001) recommended that a diet containing approximately 1.25 Mcal/kg of NEL be fed from dry off until approximately 21 d before calving, and that a diet containing 1.54 to 1.62 Mcal/kg of NEL be fed during the last 3 wk preceding parturition.

The primary rationale for feeding a lower energy diet during the early dry period is to minimizeBCSgain during the dry period; furthermore, Dann et al. (2003) reported recently that supplying excessive energy to dairy cows during the early dry period may actually have detrimental carryover effects during the subsequent early lactation period. The nature of these carryover effects is not known. One could speculate, however, that effects could be mediated through metabolic machinery responsible for tissue responsiveness to endocrine signals during the late prepartum period. In general, available information supports feeding the higher energy diet for two to three weeks prior to parturition (Mashek and Beede, 2001; Corbett, 2002; Contreras et al., 2004). Results from 2 of these experiments indicated farm-specific negative effects on subsequent production and health if cows were fed the higher energy diet for the entire dry period (Contreras et al., 2004) or for an average of 37 d prepartum (Mashek and Beede, 2001). These responses may correspond to the negative carryover effects of overfeeding energy during the early dry period described by Dann et al. (2003).

Furthermore, recent results (Contreras et al., 2004) support managing cows to achieve a BCS of approximately 3.0 at dry off rather than the traditional 3.5 to 3.75 BCS—perhaps partially due to the decreased DMI associated with higherBCSduring the prepartum period (Hayirli et al., 2002). Studies conducted with limited replication indicate increased DMI and milk yield for cows of BCS 2 to 2.5 at calving versus those with a BCS of 3.5 to 4 on a 4-point scale.

These results are also consistent with those of Domecq et al. (1997), who reported that as BCS of cows at dry off increased, milk yield during the first 120 DIM decreased; furthermore, thinner cows that gained BCS during the dry period yielded more milk during the first 120 DIM. Collectively, results published in the scientific literature support the concept that cows of mod erately lower BCS within a well-managed transition management system are more likely to have positive transition period outcomes than cows of greater BCS due to their propensity to have increasedDMIand potentially increased milk yield during early lactation.

Strategies to Meet Glucose Demands and Decrease NEFA Supply During the Transition Period

Carbohydrate Formulation of the Prepartum Diet

A substantial amount of research has been conducted to examine carbohydrate nutrition of dairy cows during the dry period, specifically relating to the NFC content of the diet. A concept that has been perpetuated through the scientific literature (Rabelo et al., 2003) is that diets higher in NFC content than traditional dry cow diets must be fed prior to calving to promote development of ruminal papillae for adequate absorption of VFA produced during ruminal fermentation.

This idea was based on one experiment in which dry cows were adapted from a diet containing a large amount of poor-quality forage to a diet containing a much larger proportion of grain (Dirksen et al., 1985). However, Andersen et al. (1999) reported that cows fed more typical diets during the prepartum period do not have meaningful changes in ruminal epithelium.

Regardless of the effect on rumen epithelium, feeding diets containing higher proportions of NFC should promote ruminal microbial adaptation to NFC levels typical of diets fed during lactation and provide increased amounts of propionate to support hepatic gluconeogenesis and microbial protein (providing the diet contains sufficient ruminally degradable protein) to support protein requirements for maintenance, pregnancy, and mammogenesis. Although the content of both the low- and high-NFC prepartum diets varied substantially across the experiments, most of these studies reported one or more positive outcomes when the higher NFC diet was fed relative to a paired lower NFC diet. Most of the researchers reported increased prepartum DMI in response to increasing the NFC content of the prepartum diet. These results are consistent with those summarized in the correlation dataset of Hayirli et al. (2002), who reported that prepartum DMI was positively correlated with NFC content of the prepartum diet. The NFC content of the diet is only one factor that has an impact on the ruminal fermentability of carbohydrate in the diet. Accordingly, research attention also has focused on the fermentability of a given concentration of NFC in diets fed during both the prepartum and immediate postpartum periods. Dann et al. (1999) reported that increasing the fermentability of the NFC in the prepartum diet by replacing cracked corn with steam-flaked corn (39% total NFC content of the diet) tended to increase prepartum DMI, postpartum milk yield, and plasma insulin concentrations during the immediate postpartum period; NEFA concentrations were decreased during the prepartum period by increasing intake of fermentable carbohydrate.

Ordway et al. (2002) fed diets containing approximately 36% NFC during the prepartum period and replaced 2.7% of diet DM as ground shelled corn with sucrose. Feeding sucrose tended to increase plasma glucose concentrations during the prepartum period, but did not affect peripartum performance or concentrations of NEFA during either the prepartum or postpartum periods.

Most of the experiments described above confounded NFC content and energy concentration of the prepartum diet, i.e., the increase in

NFC content of the diet simultaneously increased NEL content of the diet, and, given that cows typically consumed more of the higher NFC diet, they also consumed more energy during the prepartum period. We were interested in exploring the concentration of NFC in the diet independent of energy content of the prepartum diet and, more specifically, the impact of deriving energy from starch-based NFC compared with other carbohydrate sources (Smith et al., 2002, 2003).

Of particular interest was the effect that feeding a diet high in nonforage fibre sources (i.e., beet pulp, soybean hulls) as a source of highly digestibleNDFmight have on performance and metabolism during the periparturient period. The high NFC diet contained 1.59 Mcal/kg of NEL, 40% NFC, and 28% starch; the high nonforage fibre sources diet contained 1.54 Mcal/kg of NEL, 34% NFC, and 18% starch.

Dry matter intake between the 2 treatments during the prepartum and postpartum periods did not differ, and other aspects of performance and metabolism, including the dynamics of insulin action and glucose disposal in response to a glucose challenge, were virtually unaffected by the source of dietary carbohydrate in the prepartum diet. Although we did not assess the impact of energy supply in this experiment (Smith et al., 2002, 2003); results imply that the generally positive effects on performance and metabolism of feeding diets during the prepartum period that are moderately higher in NFC content are linked to energy supply from carbohydrate rather than NFC content of the diet per se. Therefore, the specific NFC content of the diet prepartum diet may have received unwarranted focus in research and also practical diet formulation in the dairy industry. This speculation is consistent with results reported recently by Pickett et al. (2003a), who measured positive effects on metabolism and performance when NDF from forage was replaced by NDF from nonforage fibre sources in diets fed during the prepartum period.

Direct Supplementation with Glucogenic Precursors

Propylene glycol is a glucogenic precursor that has been used for many years as an oral drench in the treatment of ketosis. Available studies consistently demonstrate decreased concentrations of NEFA in plasma and usually demonstrate decreased concentrations of BHBA in plasma in response to propylene glycol administered as an oral drench. Incorporation of propylene glycol into the TMR did not affect concentrations of NEFA and BHBA in plasma (Christensen et al., 1997).

Recently, Stokes and Goff (2001) reported that administration of an oral drench of propylene glycol for 2 d beginning at parturition decreased concentrations of NEFA in plasma and increased milk yield during early lactation.

Subsequent experiments in which propylene glycol was administered as a drench beginning at parturition for either 2 (Visser et al., 2003) or 3 d (Lenkaitis et al., 2003) or as part of a combination drench administered for 3 d beginning at parturition (Visser et al., 2002) reported no productive response to propylene glycol drench. Overall, research supports that bolus administration of propylene glycol will result in modest effects on metabolic variables; however, the lack of consistent production responses across experiments dictate that routine administration of propylene glycol is not indicated.

Propionate supplements consisting of propionate complexed to Ca or trace minerals potentially could be used to supply substrate for hepatic gluconeogenesis. Published responses to peripartal supplementation with propionate supplements have been mixed. Burhans and Bell (1998) reported that postpartum supplementation of 300 g/d of Ca propionate did not affect postpartum milk yield or plasma NEFA concentrations. Mandebvu et al. (2003) reported that feeding approximately 110 g/d of a propionate supplement on a commercial dairy farm did not affect milk yield, but transiently decreased plasma NEFA concentrations and urine ketone score.

Beem et al. (2003) determined that feeding 113.5 g/d of Ca propionate during transition period did not affect DMI, milk yield, or plasma BHBA concentrations. Stokes and Goff (2001) reported that drenching cows with 0.68 kg of Ca propionate twice during the early postpartal period did not affect early lactation milk yield, or concentrations of NEFA and BHBA in plasma.

Part of the reason for the lack of measured response to propionate supplements could be the amount of propionate provided relative to the amount produced in the rumen. Midlactation cows consuming 16 kg/d of DM from a diet containing 55% forage produced almost 1000 g/d of propionate in the rumen (Bauman et al. 1971). Given that cows in the first week of lactation typically will consume a comparable amount of a comparable diet, the additional propionate supplement likely only makes a small contribution to total propionate supply to the cow. Because Stokes and Goff (2001) did not detect metabolic responses (decreased NEFA and/or BHBA concentrations in plasma) in response to oral administration of a sizable (0.68 kg) bolus of Ca propionate, we speculate that there may also be differences in either ruminal metabolism or absorption kinetics of propylene glycol and the propionate supplements. Overall, existing research does not support use of propionate supplements either through the TMR or via bolus.

Monensin provided in controlled-release capsule (CRC) form during the transition period and early lactation has been shown to decrease the incidence of subclinical ketosis in dairy cows by 50% (Duffield et al., 1998b). In addition to decreased postpartum concentrations of serum BHBA, cows administered the monensin CRC also had increased concentrations of serum glucose during the postpartum period (Duffield et al., 1998a).

Overconditioned cows (BCS > 4.0 at 21 d before expected calving) supplemented with the monensinCRCproduced significantly more milk than unsupplemented controls during early lactation (Duffield et al., 1999). In a subsequent experiment, cows administered the monensin CRC had decreased circulating concentrations of NEFA during the week immediately preceding calving; however, circulating concentrations of NEFA during the first week postcalving were not affected by administration of the monensin CRC (Duffield et al., 2003).

In contrast, cows fed 300 mg/d of monensin from 28 d prior to calving until calving did not have altered concentrations of NEFA and glucose during the prepartum period compared with controls; however, cows fed 300 mg/d of monensin during the prepartum period had significantly lower circulating NEFA concentrations during the first week postcalving (Vallimont et al., 2001).

As demonstrated with growing steers, the net effect of monensin within the rumen is to increase ruminal propionate production at the expense of ruminal acetate and methane production so that propionate supply is increased and the overall energetic efficiency of ruminal fermentation is increased (Armentano and Young, 1983). Although this mechanism is consistent with observations on the increased circulating concentrations of glucose and reduction in incidence of subclinical ketosis described above, Markantonatos et al. (2002) determined that prepartum ruminal production of propionate was not affected by feeding 300 mg/d of monensin during the prepartum period of dairy cows.

Furthermore, only modest effects of monensin on glucose kinetics during the prepartum period were measured (Arieli et al., 2001). It is uncertain whether the metabolic effect on subclinical ketosis and milk yield is mediated directly through glucose metabolism or NEFA metabolism; however, research supports consistent efficacy of monensin administered as a CRC or as a topdress. Given the fluctuations in DMI during the periparturient period, it is not known whether inclusion of monensin in a TMR will be as effective as administering via CRC. *Added fat in transition diets.* It has been proposed that dietary fat may help to decrease concentrations of NEFA and help to prevent occurrence of ketosis (Kronfeld, 1982). Dietary long-chain fatty acids are absorbed into the lymphatic system and do not pass first through the liver. This fat can provide energy for peripheral tissues and the mammary gland.

Kronfeld's hypothesis is that the increased energy availability would in turn decrease mobilization of body fat and decrease NEFA concentrations.

Despite available information, indicating that added fat fed to cows during the prepartum period does not decrease plasma NEFA concentrations, advancement of this hypothesis by various commercial in terests in the dairy industry has continued. Grum et al. (1996) determined that feeding fat (6.7% of diet DM) to cows during the entire dry period virtually abolished accumulation of triglycerides in liver during the immediate peripartal period; however, cows fed fat also had decreased DMI during the dry period. A subsequent experiment (Douglas et al., 1998) determined that the reduction in liver triglycerides was mostly attributable to the decreased DMI of cows fed added fat during the dry period. As indicated above, Doepel et al. (2002) reported that cows fed high-energy diets during the prepartum period had decreased peripartal concentrations of NEFA in plasma and tended to have decreased postpartum liver triglyceride concentrations.

The increase in energy content of the prepartum diet was achieved by a combination of increasing NFC content as reported above and addition of tallow at 2.2% of DM. The preponderance of results reported above suggest that the results of Doepel et al. (2002) occurred as a consequence of changes made in the NFC content of the diet rather than in the fat content of the diet.

Anecdotal reports from some practitioners in the dairy industry have indicated beneficial effects of administering dietary fat by oral drench to cows during the immediate postpartum period, and dietary fat sources are commonly included in commercially available mixtures administered orally to fresh cows. Pickett et al. (2003b) administered 454 g/d of a commercially available fat supplement (82% fatty acids by weight) by oral drench for the first 3 d of lactation; administration of fat did not affect concentrations of NEFA and BHBA in plasma and triglycerides in liver during the postpartum period, and tended to decrease DMI and milk yield during the first 21 d of lactation.

Effects of specific fatty acids on NEFA supply. A substantial amount of research conducted during the past few years has focused on the metabolic roles of individual fatty acids. Interest in application of individual fatty acids in transition cow nutrition and metabolism to date has focused on one of two general areas.

First, researchers have sought to determine whether feeding *trans*-10, *cis*-12 conjugated linoleic acid (CLA), a fatty acid known to decrease milk fat percentage and yield in cows in established lactation, will decrease energy output during early lactation and in turn decrease the extent and duration of negative energy balance during early lactation. Giesy et al. (1999) fed cows a mixture of CLA isomers in a Ca-salt form from d 13 through 80 postpartum.

They reported few effects of CLA supplementation on cow performance during d 14 through 28 postcalving; however, milk yield was increased, and percentage and yield of milk fat were decreased, during d 35 through 80 postpartum. Energy balance was not affected by treatment during either period. Bernal-Santos et al. (2003) fed cows a mixture of CLA isomers as Ca-salts from 14 d prepartum through 140 d postpartum. Milk fat percentage and yield decreased beginning during the third week postpartum. However, cows fed the rumen-protected CLA tended to produce more milk during early lactation and energy balance was also unaffected by treatment in this experiment.

Castaneda-Gutierrez et al. (2003) reported similar effects on milk fat percentage and yield beginning during the third week postpartum in response to feeding CLA. Milk yield was not different among treatments in this experiment. Selberg et al. (2002) fed a source of trans-octadecenoic acid during the transition period and early lactation and reported that liver triglyceride concentration decreased in response to feeding the trans-octadecenoic acid source. In contrast, Bernal-Santos et al. (2003) reported that liver triglyceride concentration was not affected by feeding CLA. Given that these 2 fatty acids appear to have different effects on liver metabolism, research is required to characterize further the metabolic effects of these 2 fatty acids in transition cows.

Nutritional Strategies to Decrease Conversion of NEFA to Accumulated Triglyceride in Liver

In addition to nutritional strategies used to decrease the supply of circulating NEFA available for extraction by the liver, the potential exists to employ nutritional strategies to decrease the rate at which NEFA are converted to triglycerides within the liver. Although hepatic capacities for disposal ofNEFAthrough mitochondrial or peroxisomal B-oxidation or export as triglycerides within VLDL are limited in ruminants compared with nonruminants (Grummer, 1993), recent evidence suggests that supplying specific nutrients to dairy cows during the transition period may increase rates of NEFA disposal, with resulting effects on performance. Choline is a quasi-vitamin that has a variety of functions in mammalian metabolism.

Its most significant functions are as a component of the predominant phospholipids contained in the membranes of all cells in the body (phosphatidylcholine), a component of the neurotransmitter acetylcholine, and as the direct precursor to betaine in methyl metabolism. Most of the potential application of choline within transition cow nutrition has focused on its role in lipid metabolism because phosphatidylcholine is required for synthesis and release of VLDL by liver. Choline deficiency in rats resulted in a sixfold increase in liver triglyceride content (Yao and Vance, 1990), and in vitro incubation of hepatocytes isolated from choline-deficient rats with either choline or Met increased concentrations of phosphatidylcholine in liver and release of VLDL (Yao and Vance, 1988).

Feed ing choline in rumen-protected form to transition dairy cows tended to decrease the rate of accumulation of esterified products in liver slices in vitro, implying that VLDL export was sensitive to choline supply also in dairy cows. Yields of milk and fat-corrected milk have generally increased in response to feeding rumen-protected choline during the transition period, suggesting that the metabolic changes in hepatic fatty acid metabolism translated into improved performance during early lactation.

Methionine and Lys are frequently considered to be the 2 most limiting AA for synthesis of milk and milk protein (NRC, 2001). These 2 AA also have potential roles in mitochondrial B-oxidation of fatty acids (carnitine biosynthesis) in liver and export of triglycerides as VLDL (apolipoprotein B100 biosynthesis; Bauchart et al., 1998). A potential role for Met in bovine ketosis has been speculated for more than 30 yr. Investigators that have sought to increase the supply of Met as either rumen- protected Met (Socha et al., 1994; Overton et al., 1996) or its analog [(2-hydroxy-4-(methylthio)-butanoic acid; Rode et al., 1998; Piepenbrink et al., 2004] beginning prior to parturition and continuing through early lactation generally reported increased milk yield during early lactation.

These positive productive responses do not appear to relate directly to effects of Met and Lys on hepatic lipid or glucose metabolism. Therefore, specific roles for Met and Lys in aspects of hepatic lipid and glucose metabolism remain speculative and unsubstantiated. Linoleic and linolenic acids are considered to be essential in many species. Linolenic acid is a precursor to both docosahexaenoic and eicosapentaenoic acids—collectively, these fatty acids may have roles important for the secretion of apolipoprotein B100 and also for VLDL particle stability in cultured hepatocytes. Consistent with these effects, incubation of ruminant hepatocytes in vitro demonstrated a potential role of linolenic acid in decreasing cellular accumulation of triglycerides from palmitic acid (Mashek et al., 2002). Short-term cultures of liver slices from immediate postpartal cows displayed decreased capacity for fatty acid esterification when incubated with a mixture of linoleic and linolenic acids (Piepenbrink and Overton, 2003a). These initial results are intriguing— are the effects of linolenic acid and its products specific to VLDL export, or are there potential effects of these fatty acids on either mitochondrial or peroxisomal B-oxidation?

Restricted Feeding During the Dry Period

Despite the widely held concept that increased DMI during the prepartum period is a harbinger of increased DMI during the postpartum period and overall transition cow success, several investigators have studied the potential to restrict energy intake of dairy cows during the prepartum period to precondition metabolism to negative energy balance. In general, cows fed balanced diets restricted to below calculated energy requirements (usually about 80% of predicted requirements) did not decrease their voluntary DMI during the days preceding parturition and increased postpartum DMI and milk yield at faster rates than cows consuming the same diets for ad libitum intake.

Furthermore, feed-restricted cows typically had blunted peripartal NEFA curves compared with those fed for ad libitum intake, and cows fed for ad libitum intake prepartum had decreased insulin sensitivity compared with those that were restricted-fed (Holtenius et al., 2003). Collectively these results are intriguing, but all experiments were conducted with cows that were individually fed. Achieving uniform restricted intake in the typical group-fed situation on commercial farms will be difficult to achieve. The phenomena of improved health and performance when DMI of cows is restricted during the prepartum period so that voluntary DMI prior to calving is not decreased has led to increasing focus on the dynamics of the prepartum DMI curve (i.e., the rate and extent of decrease of DMI prior to parturition). Recently, Mashek and Grummer (2003) proposed that, although postpartum DMI and milk production appeared to correlate more strongly with total DMI from 21 d prepartum to 1 d prepartum, the change in DMI from 21 d prepartum to 1 d prepartum correlated more strongly with metabolic indices such as postpartum plasma NEFA concentrations and liver triglyceride accumulation. The metabolic events underpinning these relationships are not known. It is possible that tissue specific responses to endocrine signals may be affected by plane of nutrition during the prepartum period, given that overall concentrations of hormones typically are only modestly affected by dietary treatment during this timeframe.

Considerations for Prevention of Hypocalcemia

One of the more interesting relationships in the epidemiological study of Curtis et al. (1985) was the lack of association of Ca content of the diet fed prepartum with occurrence of milk fever. Indeed, the NRC (2001) effectively discounted the potential that diets sufficiently low in Ca to prevent hypocalcemia could be fed during the prepartum period.

In turn, they focused attention on the approach of adjusting cation-anion difference [[Na++K+] [Cl" + S"2]] to prevent metabolic alkalosis and perhaps induce a compensated metabolic acidosis. Horst et al. (1997) hypothesized that this correction of metabolic alkalosis would prevent changes in the conformation of the receptor for parathyroid hormone on bone and facilitate mobilization of Ca from bone. Prepartal diets with a negative dietary cation-anion difference (DCAD) have repeatedly been shown to reduce subclinical and clinical hypocalcemia in cows predisposed to milk fever (Horst et al., 1997). Although the basic tenet of DCAD has not changed since their review, this area has been the subject of much research, and several aspects of this subject deserve mention.

The use of prepartum diets having a lower DCAD has repeatedly been shown to be effective in preventing milk fever in cows predisposed to milk fever. Nonetheless, several points should be noted that should affect decision making for feeding low DCAD diets. As pointed out by Roche et al. (2002), much of the controlled research concerning diets containing a decreased DCAD was conducted using animals that are highly predisposed to milk fever (e.g., Jersey cows in their third or greater lactation). Therefore, the effects of a low DCAD diet in breeds that are less susceptible to milk fever and in modern dairy herds that often exceed 40% first-lactation animals need to be considered in decision-making.

Moore et al. (2000) reported that although inclusion of anionic salts in the diet of firstlactation Holstein cows effectively induced a compensated metabolic acidosis, Ca metabolism was not improved, prepartum DMI was reduced, prepartum circulating NEFA concentrations were increased, and more triglycerides accumulated in the liver. Furthermore, these authors reported anecdotally that compared to a control cow that twinned, cows carrying twins fed anionic salts to "15 meq/100 g of diet had dramatically decreased DMI and increased circulatingNEFA and liver triglyceride concentrations.

Currently, controversy exists regarding whether sufficient alleviation of hypocalcemia can occur by decreasing the cation (Na and K) content of the diet fed during the prepartum period alone without adding anions through mineral- or acid- (HCl) based sources. Existing research in the literature is equivocal about whether a reduction in dietary K and a moderate DCAD are sufficient to avert milk fever or whether herds predisposed to milk fever might benefit from diets with a DCAD of “10 to “15 meq/100 g of DM.

Goff and Horst (1997) indicated a reduction in dietary K to 1.1% DM was sufficient to avert clinical milk fever in multiparous Jersey cows; however, the incidence of subclinical hypocalcemia was not reduced. Moore et al. (2000) reported that cows fed a diet containing aDCAD of 0 meq/100 gDMresulted in intermediate indices of Ca metabolism relative to cows fed diets containing either “15 or +15 meq/ 100 g DM; however, the feeding a diet with a 0 DCAD was not sufficient to prevent parturient hypocalcemia in Holstein cows.

A final area of research that does not directly fit into DCAD programs but has stimulated interest in the area of Ca metabolism is the inclusion of the sodium aluminum silicate Zeolite A in the diets of dairy cattle. Negative Ca balance induces Ca homeostatic mechanisms to be upregulated such that the severe decline in serum Ca at the onset of lactation is attenuated (Horst et al., 1997). Indeed restriction of prepartum Ca intake to less than 20 g/d has proven effective in preventing parturient paresis; however, this dramatic restriction of dietary Ca is difficult to achieve with available feedstuffs.

Zeolite A potentially binds Ca in the digestive tract thereby making it unavailable for intestinal absorption by the cow and, in theory, dramatically restricts Ca entry rate to induce negative Ca balance in the cow prior to the initiation of lactogenesis. Thilsing-Hansen and Jorgensen (2001) reported that dietary supplementation with Zeolite A prepartum prevented milk fever and subclinical hypocalcemia in Jersey cows. Similarly, Thilsing- Hansen et al. (2002) reported that although control cows were not hypocalcemic, cows fed ZeoliteAprepartum had higher serum concentrations of 1,25-dihydroxyvitaminD about 1 wk prior to calving and had greater serum Ca levels on the day of calving. Although these initial results are promising, more research is needed on this supplement to determine actual in vivo Ca binding capacity and also to determine any potential negative effects on the bioavailability of micronutrients when cows are fed Zeolite A.

Chapter 11

Poultry Farming

Poultry is a category of domesticated birds kept by humans for the purpose of collecting their eggs, or killing for their meat and/or feathers. These most typically are members of the superorder Galloanserae (fowl), especially the order Galliformes (which includes chickens, quails and turkeys) and the family Anatidae (in order Anseriformes), commonly known as "waterfowl" (e.g. domestic ducks and domestic geese). Poultry also includes other birds which are killed for their meat, such as pigeons or doves or birds considered to be game, like pheasants. Poultry comes from the French/Norman word, poule, itself derived from the Latin word Pullus, which means small animal.

Cuts of Poultry

The meatiest parts of a bird are the flight muscles on its chest, called breast meat, and the walking muscles on the first and second segments of its legs, called the thigh and drumstick, respectively. Dark meat, which avian myologists (bird muscle scientists) refer to as "red muscle," is used for sustained activity—chiefly walking, in the case of a chicken. The dark color comes from the protien myoglobin, which plays a key role in oxygen uptake within cells. White muscle, in contrast, is suitable only for short, ineffectual bursts of activity such as, for chickens, flying. Thus the chicken's leg and thigh meat are dark while its breast meat (which makes up the primary flight muscles) is white. Other birds with breast muscle more suitable for sustained flight, such as ducks and geese, have red muscle (and therefore dark meat) throughout.

Chicken (food): Chicken is the fowl derived from chickens. It is the most common type of poultry in the world, and is prepared as food in a wide variety of ways, varying by region and culture.

History

The modern chicken is a descendant of Red Junglefowl hybrids along with the Grey Junglefowl first raised thousands of years ago in the northern parts of the Indian subcontinent.

Chicken as a meat has been depicted in Babylonian carvings from around 600 BC. Chicken was one of the most common meats available in the Middle Ages. It was widely believed to be easily digested and considered to be one of the most neutral foodstuff. It was eaten over most of the Eastern hemisphere and a number of different kinds of chicken such as capons, pullets and hens were eaten. It was one of the basic ingredients in the so-called white dish, a stew usually consisting of chicken and fried onions cooked in milk and seasoned with spices and sugar.

Chicken consumption in the US increased during World War II due to a shortage of beef and pork. In Europe, consumption of chicken overtook that of beef and veal in 1996, linked to consumer awareness of Bovine spongiform encephalopathy or B.S.E.

Breeding

Modern varieties of chicken such as the Cornish Cross, are bred specifically for meat production, with an emphasis placed on the ratio of feed to meat produced by the animal. The most common breeds of chicken consumed in the US are Cornish and White Rock.

Chickens raised specifically for food are called broilers. In the United States, broilers are typically butchered at a young age. Modern Cornish Cross hybrids, for example, are butchered as early as 8 weeks for fryers and 12 weeks for roasting birds.

Capons (castrated cocks) produce more and fattier meat. For this reason, they are considered a delicacy and were particularly popular in the [Middle Ages].

Edible Components

Typically, the muscle tissue (breast, legs, thigh, etc.), liver, heart, and gizzard are processed for food. Chicken feet are commonly eaten, especially in Caribbean and Chinese cuisine. Chicken wings refers to a serving of the wing sections of a chicken.

Exotic parts like pygostyle (chicken's buttocks) and testicles are commonly eaten in East Asia and some parts of South East Asia.

Chicken eggs are commonly eaten. Chicken is considered poultry rather than meat.

Health Issues

Chicken meat is generally considered healthy (in contrast to red meat), however, due to a significant inclusion of cholesterol (as in most meats) it should be avoided in excess, especially for conditions in sensitivity of it (the exact conditions around cholesterol intake from animal products (limit, effects, etc.) is debated). The cholesterol issue is also apparent on the animal's eggs (particularly the yolk), and is more pronounced in certain parts of the body.

Chicken generally includes low fat in the meat itself (castrated roosters excluded), however it is highly concentrated on its skin, which should be avoided when low intake of fat is necessary.

Cooking method is highly related to health issues related to chicken, with steam cooking with skin removed being considered one of the healthiest, and fried with trans fats one of the worst.

However according to a 2006 Harvard School of Public Health study of 135,000 people, people who ate grilled skinless chicken 5 or more times a week had a 52 percent higher chance of developing bladder cancer compared to people who didn't. However, such strong associations were not found in individuals regularly consuming chicken with skin intact.

Marketing and Sales

Juvenile chickens, of less than 28 days of age at slaughter in the United Kingdom are marketed as poussin. Mature chicken is sold as small, medium or large.

Whole mature chickens are marketed in the United States as fryers, broilers, and roasters. Fryers are the smallest size, and the most common, as chicken reach this size quickly. Most dismembered packaged chicken would be sold whole as fryers. Broilers are larger than fryers. They are typically sold whole. Roasters, or roasting hens, are the largest chickens sold and are typically more expensive. These names reflect the most appropriate cooking method for the surface area to volume ratio. As the size increases, the volume (which determines how much heat must enter the bird for it to be cooked) increases faster than the surface area (which determines how fast heat can enter the bird). For a fast method of cooking, such as frying, a small bird is appropriate: frying a large piece of chicken results in the inside being undercooked when the outside is ready.

Chicken is also sold in dismembered pieces. Pieces may include quarters, or fourths of the chicken. A chicken is typically cut into two leg quarters and two breast quarters. Each quarter contains two of the commonly available pieces of chicken. A leg quarter contains the thigh, drumstick and a portion of the back; a leg has the back portion removed. A breast quarter contains the breast, wing and portion of the back; a breast has the back portion and wing removed. Pieces may be sold in packages of all of the same pieces, or in combination packages. Whole chicken cut up refers to either the entire bird cut into 8 individual pieces. (8-piece cut); or sometimes without the back. A 9-piece cut (usually for fast food restaurants) has the tip of the breast cut off before splitting. Pick of the Chicken, or similar titles, refers to a package with only some of the chicken pieces. Typically the breasts, thighs, and legs without wings or back.

Thighs and breasts are sold boneless and/or skinless. Dark meat (legs, drumsticks and thighs) pieces are typically cheaper than white meat pieces (breast, wings). Chicken livers and/or gizzards are commonly available packaged separately. Other parts of the chicken, such as the neck, feet, combs, etc. are not widely available except in countries where they are in demand, or in cities that cater to ethnic groups who favor these parts.

There are many fast food restaurant chains on both a national and global scale that sell exclusively or primarily in poultry products including KFC (global), Red Rooster (Australia), Hector Chicken (Belgium) and CFC (Indonesia). Most of the products on the menu in such eateries are fried or breaded and are served with french fries.

Use of Roxarsone in Chicken Production

In many factory farms, chickens are routinely administered with the feed additive Roxarsone, a relatively benign organoarsenic compound which partially decomposes into inorganic arsenic compounds in the flesh of chickens, and in their feces, which are often used as a fertilizer. The compound is used to control stomach pathogens and promote growth.

A *Consumer Reports* study in 2004 reported finding "no detectable arsenic in our samples of muscle" but found "A few of our chicken-liver samples has an amount that according to EPA standards could cause neurological problems in a child who ate 2 ounces of cooked liver per week or in an adult who ate 5.5 ounces per week." However, the amounts found in these livers averaged to 460 part per billion; an amount still less than the 2,000 parts per billion limit set by the FDA.

The FDA has found there to be no significant impact for the use of Roxarsone in the production of chicken, turkey or swine.

Duck (food): Duck refers to the meat of several species of bird in the Anatidae family, found in both fresh and salt water. Duck is eaten in many cuisines around the world.

Types of Ducks

The most common duck meat consumed in the United States is the Pekin duck. Because most commercially raised Pekins come from Long Island, New York, Pekins are also sometimes called "Long Island" ducks, despite being of Chinese origin. Some specialty breeds have become more popular in recent years, notably the Muscovy duck, and the Moulard duck (a sterile hybrid of Pekins and Muscovies). Unlike most other domesticated ducks, Muscovy ducks are not descended from mallards.

According to the USDA, nearly 26 million ducks were eaten in the U.S. in 2004.

Duck Meat

Duck meat is derived primarily from the breasts and legs of ducks. The meat of the legs is darker and somewhat fattier than the meat of the breasts, although the breast meat is darker than the breast meat of a chicken or a turkey. Being waterfowl, ducks have a layer of heat-insulating subcutaneous fat between the skin and the meat. De-boned duck breast can be grilled like steak, usually leaving the skin and fat on. Magret refers specifically to the breast of a mallard or Barbary duck that has been force fed to produce foie gras.

Internal organs such as heart and kidneys may also be eaten; the liver in particular is often used as a substitute for goose liver in foie gras.

Dishes

Duck is used in a variety of dishes around the world, most of which involve roasting for at least part of the cooking process to aid in crisping the skin. Notable duck dishes include:

- Bebek Betutu: a famous traditional dish from Bali, Indonesia. The duck is first seasoned with pungent roots and various herbs, wrapped with banana leaves, and roasted. Chicken is also used to prepare Betutu.
- Confit: duck legs that have been cured (partly or fully) in salt, then marinated and poached in duck fat, typically with garlic and other herbs. The French word *confit* means "preserved", and the French name for duck confit is "confit de canard."

- Czernina: a sweet and sour Polish soup made of duck blood and clear poultry broth. It was once considered a symbol of Polish culture until the 19th century, customarily served to young men and is even featured as a plot device in a famous epic poem called Pan Tadeusz.
- Duck à l'orange: a classic French dish in which the duck is roasted and served with an orange sauce.
- Duck sauce : a fruit-based dipping sauce for deep fried dishes in Chinese American cuisine.
- Foie Gras: a specially fattened and rich liver, or a pâté made from the liver, sometimes taken from a duck but usually from a goose.
- Oritang: a variety of *guk*, Korean soup made with duck and various vegetables.
- Peking Duck: a famous Chinese dish originating from Beijing, prepared since the Ming Dynasty era. It is prized for the thin, crispy skin, with authentic versions of the dish serving mostly the skin and little meat, and eaten with pancakes, spring onions, and hoisin sauce or sweet bean sauce.
- Pressed duck: a complex dish originally from Rouen, France.
- Turducken: an American dish that comprises a turkey, stuffed with a duck, which is in turn stuffed with a chicken.
- Zhangcha duck: a quintessential dish of Sichuan cuisine. It is first prepared by smoking a marinated duck over tea leaves and twigs of the camphor plant, then steamed, and finally deep fried for a crisp finish. Also called tea-smoked duck.
- Long Island roast duckling: This is a whole roasted bird, sometimes brined previously. When done properly, most of the fat melts off during the cooking process, leaving a crispy skin and well-done meat. Some restaurants on Long Island serve this dish with a cherry sauce.

Oyster (Fowl)

Oysters are two small, round pieces of dark meat on the back of poultry near the thigh, in the hollow on the dorsal side of the ilium bone. Some regard the "oyster meat" to be the most flavourful and tender part of the bird, while others dislike the taste and texture.

Compared to dark meat found in other parts of the bird, the oyster meat has a somewhat firm/taut texture which gives it a distinct mouth feel.

A common misconception is that the oyster is the gluteus maximus muscle of the bird.

In French, this part of the bird is called "sot-l'y-laisse" which translates, roughly, to "the fool leaves it there", as unskilled carvers sometimes accidentally leave it on the skeleton.

Poultry Farming

Poultry farming is the practice of raising poultry, (also known as such as raggle fraggle) chickens, turkeys, ducks, and geese, as a subcategory of animal husbandry, for the purpose of farming meat or eggs for food.

More than 50 billion chickens are reared annually as a source of food, for both their meat and their eggs. Chickens farmed for meat are called broilers, whilst those farmed for eggs are called egg-laying hens. In total, the UK alone consumes over 29 million eggs per day. Some hens can produce over 300 eggs a year. Chickens will naturally live for 6 or more years. After 12 months, the hen's productivity will start to decline. This is when most commercial laying hens are slaughtered.

The majority of poultry are raised using intensive farming techniques. According to the Worldwatch Institute, 74 percent of the world's poultry meat, and 68 percent of eggs are produced this way. One alternative to intensive poultry farming is free range farming.

Friction between these two main methods has led to long term issues of ethical consumerism. Opponents of intensive farming argue that it harms the environment and creates health risks, as well as abusing the animals themselves. Advocates of intensive farming say that their highly efficient systems save land and food resources due to increased productivity, stating that the animals are looked after in state-of-the-art environmentally controlled facilities. A few countries have banned cage system housing, including Sweden and Switzerland. Consumers can still purchase lower cost eggs from other countries' intensive poultry farms.

Techniques

Free-range

Free range poultry farming consists of poultry permitted to roam freely instead of being contained in any manner. In the UK, the Department for Environment, Food and Rural Affairs says that a free range chicken must have daytime access to open-air runs during at least half of their life. Unlike in the United States, this definition also applies to eggs. The European Union regulates marketing standards for egg farming which specifies a minimum condition for Free Range Eggs states that "hens have continuous daytime access to open-air runs, except in the case of temporary restrictions imposed by veterinary authorities". In free-range broiler systems, the chickens are given continuous access to an outdoor range during the daytime and sheds where they are housed at night. Free-range chickens grow more slowly than intensive chickens. They live at least 56 days. In the EU each chicken must have one square metre of outdoor space.

Free range poultry production requires that the poultry have access to the outside. In some cases this means the poultry are raised on pasture, enabling the poultry to move around, forage for their natural diet and live in cleaner conditions than those in batteries. In some farms, the manure from free range poultry can be used to benefit crops.

The benefits are also a reduced growth rate and opportunities for natural behaviour such as pecking, scratching, foraging and exercise outdoors, as well as fresh air and daylight. Because they grow slower and have opportunities for exercise free-range chickens have better leg and heart health and a much higher quality of life.

Finding suitable land with adequate drainage to minimise worms and coccidial oocysts, suitable protection from prevailing winds, good ventilation, access and protection from predators can be difficult. Excess heat, cold or damp can have a harmful effect on the animals and their productivity. Unlike battery farms, free range farmers have little control over the food their animals come across which can lead to unreliable productivity.

Some free range farming in the UK, which accounts for 26% of production, has also come under criticism concerning animal welfare. This is due to some large scale free range farms where social abnormalities arise due to having large numbers of birds in an outdoor space. Beak trimming due to cannibalism and infighting is common in this form of poultry farming as well as in batteries. Diseases are common and the animals are vulnerable to predators. In South-East Asia, a lack of disease control in free range farming has been associated with outbreaks of Avian influenza.

In organic systems, chickens are also free-range. Organic chickens are slower growing, more traditional breeds and live typically for around 81 days. They grow at half the rate of intensive chickens. They have a larger space allowance outside (at least 2 square metres and sometimes up to 10 square metres per bird).

Yarding

While often confused with free-range farming, yarding is actually a separate method of poultry culture by which chickens and cows are raised together. The distinction is that free-range poultry are either totally unfenced, or the fence is so distant that it has little influence on their freedom of movement. Yarding is common technique used by small farms in the Northeastern US.

Daily releases out of hutches or coops allows for instinctual nature for the chickens with protections from predators. The hens usually lay eggs either on the ground of the coop or in baskets if provided by the farmer. This technique can be complicated if used with roosters though, mostly because of difficulty getting them into the coop and to clean the coop while it is inside. This territorial nature is apparent while outside in which they have a brood of hens and sometimes even informal land claims. This can endanger people unaware of the existence of the territories who are attacked by the larger birds.

Intensive Chicken Farming

In egg-producing farms, birds are typically housed in rows of battery cages. Environmental conditions are automatically controlled, including light duration, which mimics summer daylength. This stimulates the birds to continue to lay eggs all year round. Normally, significant egg production only occurs in the warmer months. Critics argue that year-round egg production stresses the birds more than normal seasonal production.

Meat chickens, commonly called broilers, are floor-raised on litter such as wood shavings or rice hulls, indoors in climate-controlled housing. Poultry producers routinely use nationally approved medications, such as antibiotics, in feed or drinking water, to treat disease or to prevent disease outbreaks arising from overcrowded or unsanitary conditions. In the U.S., the national organization overseeing chicken production is the Food and Drug Administration (F.D.A.). Some F.D.A.-approved medications are also approved for improved feed utilization.

In egg-producing farms, cages allow for more birds per unit area, and this allows for greater productivity and lower space and food costs, with more efforts put into egg-laying. In the U.S., for example, the current recommendation by the United Egg Producers is 67 to 86 in^2 (430 to 560 cm^2) per bird, which is about 9 inches by 9 inches. Modern poultry farming is very efficient and allows meat and eggs to be available to the consumer in all seasons at a lower cost than free range production, and the poultry have no exposure to predators.

The cage environment of egg producing does not permit birds to roam. The closeness of chickens to one another frequently causes cannibalism. Cannibalism is controlled by de-beaking (removing a portion of the bird's beak with a hot blade so the bird cannot effectively peck). Another condition that can occur in prolific egg laying breeds is osteoporosis. This is caused from year-round rather than seasonal egg production, and results in chickens whose legs cannot support them and so can no longer walk. During egg production, large amounts of calcium are transferred from bones to create eggshell. Although dietary calcium levels are adequate, absorption of dietary calcium is not always sufficient, given the intensity of production, to fully replenish bone calcium.

Under intensive farming methods, a meat chicken will live less than six weeks before slaughter. This is half the time it would take traditionally. This compares with free-range chickens which will usually be slaughtered at 8 weeks, and organic ones at around 12 weeks.

In intensive broiler sheds, the air can become highly polluted with ammonia from the droppings. This can damage the chickens' eyes and respiratory systems and can cause painful burns on their legs (called hock burns) and feet. Chickens bred for fast growth have a high rate of leg deformities because they cannot support their increased body weight. Because they cannot move easily, the chickens are not able to adjust their environment to avoid heat, cold or dirt as they would in natural conditions. The added weight and overcrowding also puts a strain on their hearts and lungs. In the U.K., up to 19 million chickens die in their sheds from heart failure each year.

Indoor with Higher Welfare

Chickens are kept indoors but with more space (around 12 to 14 birds per square metre). They have a richer environment for example with natural light or straw bales that encourage foraging and perching. The chickens grow more slowly and live for up to two weeks longer than intensively farmed birds. The benefits of higher welfare indoor systems are the reduced growth rate, less crowding and more opportunities for natural behaviour.

Issues with Poultry Farming

Humane Treatment

Animal welfare groups have frequently criticized the poultry industry for engaging in practices which they believe to be inhumane. Many animal rights advocates object to killing chickens for food, the "factory farm conditions" under which they are raised, methods of transport, and slaughter. Compassion Over Killing and other groups have repeatedly conducted undercover investigations at chicken farms and slaughterhouses which they allege confirm their claims of cruelty.

Conditions in intensive chicken farms may be unsanitary, allowing the proliferation of diseases such as salmonella and E. coli. Chickens may be raised in total darkness; hens are most often kept in crowded wire battery cages with space less than that of a sheet of paper per hen, as opposed to cage-free or free range. Rough handling and crowded transport during various weather conditions and the failure of existing stunning systems to render the birds unconscious before slaughter have also been cited as welfare concerns. Another animal welfare concern is the use of selective breeding to create heavy, large-breasted birds, which can lead to crippling leg disorders and heart failure for some of the birds. Concerns have been raised that companies growing single varieties of birds for eggs or meat are increasing their susceptibility to disease. A common practice among hatcheries is the culling of newly born male chicks of egg laying breeds, since they don't lay eggs, and do not grow fast enough to be profitable for meat.

Debeaking

Laying hens are routinely de-beaked when young to prevent fighting. This is bad because beaks are sensitive, the usual practice of trimming them without anaesthesia is considered inhumane by some. De-beaked chickens will peck much less than chickens with beaks, which animal behaviorist Temple Grandin attributes to guarding against pain. The United Egg Producers says that de-beaking is not painful. It is also argued that the procedure causes life-long chronic pain and discomfort and decreased ability to eat or drink.

Intelligence

Some groups which advocate for more humane treatment of chickens claim that chickens are intelligent. Dr. Chris Evans of Macquarie University claims that their range of 20 calls, problem solving skills, use of representational signaling, and the ability to recognize each other by facial features demonstrate the intelligence of chickens.

Destructive Practices that Pose Serious Risk to Humans

Antibiotics

Antibiotics have been used on poultry in large quantities since the 1940s, when it was found that the byproducts of antibiotic production, fed because the antibiotic-producing mold had a high level of vitamin B_{12} after the antibiotics were removed, produced higher growth than could be accounted for by the vitamin B_{12} alone. Eventually it was discovered that the trace amounts of antibiotics remaining in the byproducts accounted for this growth.

The mechanism is apparently the adjustment of intestinal flora, favoring "good" bacteria while suppressing "bad" bacteria, and thus the goal of antibiotics as a growth promoter is the same as for probiotics. Because the antibiotics used are not absorbed by the gut, they do not put antibiotics into the meat or eggs.

Antibiotics are used routinely in poultry for this reason, and also to prevent and treat disease. Many contend that this puts humans at risk as bacterial strains develop stronger and stronger resistances. Critics point out that, after six decades of heavy agricultural use of antibiotics, opponents of antibiotics must still make arguments about theoretical risks, since actual examples are hard to come by. Those antibiotic-resistant strains of human diseases whose origin is known originated in hospitals rather than farms.

A proposed bill in the United States Congress would make the use of antibiotics in animal feed legal only for therapeutic (rather than preventative) use, but it has not been passed. However, this may present the risk of slaughtered chickens harboring pathogenic bacteria and passing them on to humans that consume them.

In October 2000, the U.S. Food and Drug Administration (FDA) discovered that two antibiotics were no longer effective in treating diseases found in factory-farmed chickens; one antibiotic was swiftly pulled from the market, but the other, Baytril, was not. Bayer, the company which produced it, contested the claim and as a result, Baytril remained in use until July 2005.

Arsenic

Chicken feed can also include Roxarsone, an antimicrobial drug that also promotes growth. Roxarsone was used as a broiler starter by about 70% of the broiler growers between 1995 to 2000. The drug has generated controversy because it contains arsenic, which is highly toxic to humans. This arsenic could be transmitted through run-off from the poultry yards. A 2004 study by the U.S. magazine Consumer Reports reported "no detectable arsenic in our samples of muscle" but found "A few of our chicken-liver samples has an amount that according to EPA standards could cause neurological problems in a child who ate 2 ounces of cooked liver per week or in an adult who ate 5.5 ounces per week." The U.S. Food and Drug Administration (FDA), however, is the organization responsible for the regulation of foods in America, and all samples tested were "far less than the... amount allowed in a food product."

Growth Hormones

Chickens grow much more rapidly than they once did and some consumers have concluded that this rapid growth is due to the use of hormones in these animals. Some consumers believe that the increasingly earlier onset of puberty in humans is the result of the liberal use of such hormones. However, hormone use in poultry production is illegal in the United States. Similarly, no chicken meat for sale in Australia is fed hormones. Furthermore, several scientific studies have documented the fact that chickens grow rapidly because they are bred to do so. A small producer of natural and organic chickens confirmed this assumption:

" If this were 1948, you might have something to worry about. Using hormones to boost egg production was a brief fad in the Forties, but was abandoned because it didn't work. Using hormones to produce soft-meated roasters was used to some extent in the Forties and Fifties, but the increased growth rates of broilers made the practice irrelevant—the broilers got as big as anyone wanted them to get when they were still young enough to be soft-meated without chemicals.

The only hormone that was ever used in any quantity on poultry (DES) was banned in 1959, after everyone but a few die-hard farmers had given them up as a silly idea. Hormones are now illegal in poultry and eggs. The people who advertise "No hormones" are either woefully ignorant or are indulging in cynical fear-mongering, maybe both."

E. coli

According to Consumer Reports, "1.1 million or more Americans sickened each year by undercooked, tainted chicken." A USDA study discovered *E. coli* in 99% of supermarket chicken, the result of chicken butchering not being a sterile process. However, the same study also cautions that the type of *E. coli* turned up was in every case a non-lethal form distinct from the more dangerous "O157:H7" strain. Many of these chickens, furthermore, had relatively low levels of contamination. Feces tend to leak from the carcass until the evisceration stage, and the evisceration stage itself gives an opportunity for the interior of the carcass to receive intestinal bacteria. (So does the skin of the carcass, but the skin presents a better barrier to bacteria and reaches higher temperatures during cooking). Before 1950, this was contained largely by not eviscerating the carcass at the time of butchering, deferring this until the time of retail sale or in the home. This gave the intestinal bacteria less opportunity to colonize the edible meat. The development of the "ready-to-cook broiler" in the 1950s added convenience while introducing risk, under the assumption that end-to-end refrigeration and thorough cooking would provide adequate protection. *E. coli* can be killed by proper cooking times, but there is still some risk associated with it, and its near-ubiquity in commercially farmed chicken is troubling to some. Irradiation has been proposed as a means of sterilizing chicken meat after butchering.

Avian Influenza

There is also a risk that crowded conditions in chicken farms will allow avian influenza (bird flu) to spread quickly. A United Nations press release states: "Governments, local authorities and international agencies need to take a greatly increased role in combating the role of factory-farming, commerce in live poultry, and wildlife markets which provide ideal conditions for the virus to spread and mutate into a more dangerous form..."

Efficiency

Farming of chickens on an industrial scale relies largely on high protein feeds derived from soybeans; in the European Union the soybean dominates the protein supply for animal feed, and the poultry industry is the largest consumer of such feed. Two kilograms of grain must be fed to poultry to produce 1 kg of weight gain. However, for every gram of protein consumed, chickens yield only 0.33 g of edible protein.

World Chicken Population

The Food and Agriculture Organization of the United Nations estimated that in 2002 there were nearly sixteen billion chickens in the world, counting a total population of 15,853,900,000. The figures from the *Global Livestock Production and Health Atlas* for 2004 were as follows:

1.	China	(3,860,000,000)
2.	United States	(1,970,000,000)
3.	Indonesia	(1,200,000,000)
4.	Brazil	(1,100,000,000)
5.	Mexico	(540,000,000)
6.	India	(425,000,000)
7.	Russia	(340,000,000)
8.	Japan	(286,000,000)
9.	Iran	(280,000,000)
10.	Turkey	(250,000,000)
11.	Nigeria	(740,000,000)

Poultry Farming in the United States

In the United States, chickens were raised primarily on family farms until about 1960. Originally, the primary value in poultry keeping was eggs, and meat was considered a byproduct of egg production. Its supply was less than the demand, and poultry was expensive. Except in hot weather, eggs can be shipped and stored without refrigeration for some time before going bad; this was important in the days before widespread refrigeration.

Farm flocks tended to be small because the hens largely fed themselves through foraging, with some supplementation of grain, scraps, and waste products from other farm ventures. Such feedstuffs were in limited supply, especially in the winter, and this tended to regulate the size of the farm flocks. Soon after poultry keeping gained the attention of agricultural researchers (around 1896), improvements in nutrition and management made poultry keeping more profitable and businesslike. Prior to about 1910, chicken was served primarily on special occasions or Sunday dinner. Poultry was shipped live or killed, plucked, and packed on ice (but not eviscerated). The "whole, ready-to-cook broiler" wasn't popular until the 1950s, when end-to-end refrigeration and sanitary practices gave consumers more confidence. Before this, poultry were often cleaned by the neighborhood butcher, though cleaning poultry at home was a commonplace kitchen skill.

Two kinds of poultry were generally offered: broilers or "spring chickens," young male chickens, a byproduct of the egg industry, which were sold when still young and tender (generally under 3 pounds live weight); and "fowls" or "stewing hens," also a byproduct of the egg industry, which were old hens past their prime for laying. This is no longer practiced; modern meat chickens are a different breed. Egg-type chicken carcasses no longer appear in stores.

The major milestone in 20th century poultry production was the discovery of Vitamin-D (named in 1922), which made it possible to keep chickens in confinement year-round. Before this, chickens did not thrive during the winter (due to lack of sunlight), and egg production, incubation, and meat production in the off-season were all very difficult, making poultry a seasonal and expensive proposition. Year-round production lowered costs, especially for broilers.

At the same time, egg production was increased by scientific breeding. After a few false starts, such as the Maine Experiment Station's failure at improving egg production, success was shown by Professor Dryden at the Oregon Experiment Station.

Improvements in production and quality were accompanied by lower labour requirements. In the 1930s through the early 1950s, 1,500 hens was considered to be a full-time job for a farm family. In the late Fifties, egg prices had fallen so dramatically that farmers typically tripled the number of hens they kept, putting three hens into what had been a single-bird cage or converting their floor-confinement houses from a single deck of roosts to triple-decker roosts. Not long after this, prices fell still further and large numbers of egg farmers left the business. This marked the beginning of the transition from family farms to larger, vertically integrated operations.

Robert Plamondon reports that the last family chicken farm in his part of Oregon, Rex Farms, had 30,000 layers and survived into the Nineties. But the standard laying house of the surviving operations is around 125,000 hens.

This fall in profitability was accompanied by a general fall in prices to the consumer, allowing poultry and eggs to lose their status as luxury foods.

The vertical integration of the egg and poultry industries was a late development, occurring after all the major technological changes had been in place for years (including the development of modern broiler rearing techniques, the adoption of the Cornish Cross broiler, the use of laying cages, etc.).

By the late 1950s, poultry production had changed dramatically. Large farms and packing plants could grow birds by the tens of thousands. Chickens could be sent to slaughterhouses for butchering and processing into prepackaged commercial products to be frozen or shipped fresh to markets or wholesalers. Meat-type chickens currently grow to market weight in six to seven weeks whereas only fifty years ago it took three times as long. This is due to genetic selection and nutritional modifications (and not the use of growth hormones, which are illegal for use in poultry in the US and many other countries). Once a meat consumed only occasionally, the common availability and lower cost has made chicken a common meat product within developed nations. Growing concerns over the cholesterol content of red meat in the 1980s and 1990s further resulted in increased consumption of chicken.

Current Status

Today, eggs are produced on large egg ranches on which environmental parameters are controlled. Chickens are exposed to artificial light cycles to stimulate egg production year-round. In addition, it is a common practice to induce molting through manipulation of light and the amount of food they receive in order to further increase egg size and production.

On average, a chicken lays one egg a day for a number of days (a "clutch"), then does not lay for one or more days, then lays another clutch. Originally, the hen presumably laid one clutch, became broody, and incubated the eggs. Selective breeding over the centuries has produced hens that lay more eggs than they can hatch. Some of this progress was ancient, but most occurred after 1900. In 1900, average egg production was 83 eggs per hen per year. In 2000, it was well over 300.

In the United States, laying hens are butchered after their second egg laying season. In Europe, they are generally butchered after a single season. The laying period begins when the hen is about 18–20 weeks old (depending on breed and season).

Males of the egg-type breeds have little commercial value at any age, and all those not used for breeding (roughly fifty percent of all egg-type chickens) are killed soon after hatching. Such "day-old chicks" are sometimes sold as food for captive and falconers birds of prey. The old hens also have little commercial value. Thus, the main sources of poultry meat a hundred years ago (spring chickens and stewing hens) have both been entirely supplanted by meat-type broiler chickens.

Traditionally, chicken production was distributed across the entire agricultural sector. In the twentieth century, it gradually moved closer to major cities to take advantage of lower shipping costs. This had the undesirable side effect of turning the chicken manure from a valuable fertilizer that could be used profitably on local farms to an unwanted byproduct. This trend may be reversing itself due to higher disposal costs on the one hand and higher fertilizer prices on the other, making farm regions attractive once more.

From the farmer's point of view, eggs used to be practically the same as currency, with general stores buying eggs for a stated price per dozen. Egg production peaks in the early spring, when farm expenses are high and income is low. On many farms, the flock was the most important source of income, though this was often not appreciated by the farmers, since the money arrived in many small payments. Eggs were a farm operation where even small children could make a valuable contribution.

Production Statistics

Recommended Culling Practices

The American Veterinary Medical Association recommends cervical dislocation and asphyxiation by carbon dioxide as the best options, but has recently amended their guidelines to include maceration, putting non-anesthetized chicks through a grinder.

The 2005-2006 American Veterinary Medical Association Executive Board held its final meeting July 13 in Honolulu, prior to the 2006 session of the House of Delegates and the AVMA Annual Convention. It proposed a policy change, which was recommended by the Animal Welfare Committee on disposal of unwanted chicks, poults, and pipped eggs. The new policy states, in part, "Unwanted chicks, poults, and pipped eggs should be killed by an acceptable humane method, such as use of a commercially designed macerator that results in instantaneous death.

Smothering unwanted chicks or poults in bags or containers is not acceptable. Pips, unwanted chicks, or poults should be killed prior to disposal. A pipped egg, or pip, is one where the chick or poult has not been successful in escaping the egg shell during the hatching process."

Poultry Farm Management

While access to the outdoors is an important feature of many alternative or free-range production systems, the indoor environment and management are also crucial. Poultry need access to an appropriate indoor environment for good production and welfare. Ideally, poultry should choose an environment; whether to be indoors or outdoors. Attention to ventilation, temperature, lighting, and litter conditions is needed. Additional good management practices include rodent control with a minimum of toxic materials.

Alternative poultry production is often on a small scale, with portable houses. Production may be certified organic. Special practices may be needed compared to conventional poultry production. Alternative poultry production is a way to boost farm income and add fertility or diversity to a farm, while providing specialty poultry products to consumers as a part of sustainable agriculture. Alternative poultry production includes systems, such as cage-free and freerange, as an alternative to conventional poultry housing and cages. Free-range systems vary widely. Some may feature large, fixed houses with yards. Others may be small portable houses regularly moved to fresh pasture. Some may consist of small shelters or pens. Alternative poultry production is often small scale, integrated onto a diversified farm, and certified organic. While free-range is a main feature, environmental conditions inside the house are also important for good welfare, particularly ventilation, temperature, lighting, and litter. A lot of information is available on environmental control in conventional poultry production.

This publication, however, focuses on alternative production, for which information is less available. Nonetheless, some practices apply to both alternative and conventional production. This publication provides "how-to" information about environment and management in alternative poultry production and highlights the practices of innovative producers. Many of the practices described here can be employed in organic production. Alternative poultry production is an important part of sustainable agriculture, protecting the environment while addressing consumer concerns.

Environment

Poultry housing should be weather-proof to provide protection from the elements (cold, rain, wind, and hot sun) and provide warmth, especially during brooding. Housing should also provide good ventilation, as well as protection from predators.

Many innovative housing designs are used in alternative poultry production, including fixed houses with permanent foundations, mobile houses, and simple shelters. For information on small-scale housing, design, materials, construction plans, ATTRA's *Range Poultry Housing.* For information on waterers, feeders, fencing, roosts, and nestboxes, ATTRA's *Poultry: Equipment for Alternative Production.* The conventional poultry industry has extensive information on large-scale housing, environmental control, and equipment that can be used for large-scale free-range or cage-free production. *Commercial Chicken Production Manual* or Extension materials. Detailed information on ventilation, lighting, and other types of environmental control are available on the University of Georgia's, Poultry House Environmental Control.

Power

While small portable houses may not use power, reliable power is important in large houses to power ventilation systems, fans, lights, heat, motors for automated feeding systems, etc. Most farms have electricity, but diesel generators can be used for power in a poultry house and are also useful for backup power in case of an electrical outage.

Temperature

The body temperature of an adult chicken is 105-107°F (40.6 to 41.7°C). The thermoneutral zone is 65-75°F (18-24°C), which allows chickens to maintain their body temperature. If the temperature is above this zone, heat must be lost in some way. Chickens have no sweat glands. Since eating increases body temperature, chickens reduce their feed intake during hot weather, and therefore gains will be less. Chickens begin panting at 85°F (29.4°C) to help dissipate heat, and drink more to avoid dehydration. A combination of high temperature and high humidity is a problem, because panting does not cool them under these conditions .

In the U.S., heat is usually more of a problem than cold. Fast-growing broilers are particularly susceptible to heat stress due to their high level of production. Producers should provide abundant cool drinking water in close proximity to the birds inside and outside. In cold months, while the conventional industry usually uses propane heaters for heating, many alternative poultry producers do not heat houses, relying only on the body warmth of the birds for heat. However, birds tend to eat more in cold temperatures, because they need more energy to stay warm outside. It may be more cost-effective to heat the house instead of paying for more feed. Heaters, such as gas brooders or heaters, can even be provided in small portable houses, with a gas tank mounted on a trailer to be moved along with the house.

To modulate temperatures, insulation under the roof is important in any climate; insulation in the walls is also helpful. Some assurance programs in Europe require that fixed houses be insulated. During hot weather, insulation keeps heat from entering, and during cold weather keeps heat from leaving the building. The greater the difference between the inside temperature and the outside temperature, the greater the need for insulation. Proper ventilation will also help regulate house temperature. Each house should have a thermometer to display the current temperature as well as the high and low temperatures in a daily period, and producers should pay attention to weather forecasts.

Ventilation

Ventilation brings fresh air into a poultry house and removes heat, moisture, and gases. Ventilation designs may be natural or mechanical. Most houses in alternative poultry production depend on natural ventilation, because doorways are usually open to provide outdoor access. There may also be additional air inlets, side curtains, or large windows that can be opened to allow more ventilation in hot weather. Ridge vents in the roof or "whirly bird" vents allow hot air to escape.

Natural ventilation makes use of the movement of air (warm air rises and cold air falls) and wind currents. A roof at least six feet tall will allow sufficient height differential for cool air to enter through low air inlets and warm air to escape through high vents. There is less control in natural ventilation than mechanical. The reasons for ventilating during winter and summer are different. During warm months, the purpose is to remove heat and control the temperature in the house, and therefore large amounts of air are moved. During cold months, the ventilation system must remove moisture and gases, especially ammonia, while conserving heat.

This is tricky because producers tend to keep houses closed up tight to conserve heat. It is done by controlling air inlets and is possible because warm air holds more moisture than cold air does. Therefore, during cold weather producers can bring small amounts of air into the house with high moisture in the air, allow the fresh air to heat to room temperature, and when this air leaves, it takes moisture out of the house. In mechanical ventilation, positive and negative pressure systems use fans to direct air into the house (positive) or exhaust air from the house (negative). The negative pressure system is most common and controls the air inlet to help mix the incoming fresh air with the air in the house. Mechanical ventilation is less appropriate for freerange houses because the doorways must be closed to maintain "static pressure."

Lighting

Poultry are very sensitive to light. Light not only allows them to be active and find their food, but it also stimulates their brains for seasonal reproduction. Light is perceived through the eyes but can also be received by other receptors in the brain, after penetrating the feathers, skin, and skull. Even blind birds respond to light. In the spectrum of visible light, blue light is relatively short wavelengths, while red light is long. Since red wavelengths are longer, they are more able to penetrate to the brain to stimulate activity and reproduction and even aggression. If the light intensity is low, then the wavelength is important.

However, if light intensity is high, then wavelength is not as important. Birds need a dark period for good health. They only produce melatonin—a hormone important in immune function— during dark periods. Welfare programs usually require at least four to six hours of dark daily, with some of the organic programs requiring eight hours of darkness. Many alternative poultry producers use only natural light and therefore have a long dark period. Dark periods can be especially helpful for fast-growing broilers in the first weeks of life to slow growth, build frame, and reduce leg disorders. (Baby chicks, however, need 24 hours of light the first three days to ensure that they learn to find food and water.)

In contrast, the conventional poultry industry uses long light periods to encourage feed consumption and weight gain by fast-growing broilers, because birds do not eat in the dark. When birds have a dark period, they are more active during the light period than birds that have continuous light. Light intensity is measured in foot-candles (fc) in the U.S. (the amount of light emitted by a standard candle at one foot away; lux is a metric measurement). For example, a brightly-lit store may be 100 fc while a home is usually 10 fc. Alternative poultry production tends to use a higher light intensity than conventional. Most welfare programs require at least 1 fc. Light intensity above 1 fc leads to increased activity, which can reduce leg problems but results in decreased weight gains.

A curtainsided house may have a light intensity of 200 fc or more when the sun is overhead, but depends on cloud cover. The conventional industry typically keeps light intensity low in poultry houses to reduce activity and gain weight more efficiently. The conventional industry uses about 0.5 fc or less, similar to a moonlit night, for broilers and layers. Both conventional and alternative egg producers use artificial lighting to stimulate production during days of declining natural light, resulting in a more constant supply of eggs. Small-scale producers often use 14 hours of light for layers. Generally the light period should not be longer than the longest day of the year. Day length should not be increased for young growing pullets or they will begin producing eggs too soon; likewise, day length should not be decreased for layers and breeders in production or they will stop producing eggs. Sunlight is a broad spectrum white light and contains all the wavelengths of visible light. Common types of artificial light are incandescent and fluorescent. Incandescent lights are a broad spectrum light with a predominance of long (red) wavelengths. Fluorescent lights are a variable light spectrum, depending on their manufacture.

Two types commonly used in poultry houses are "warm white" and "cool white." Warm white has a predominance of long yellow wavelengths, and cool white has a predominance of shorter blue to green wavelengths. Fluorescent lights come in tube and compact forms.

The 2700 K compact is similar to the "warm white" and the 2700 K is similar to the "cool white." Incandescent lights are easier and less expensive to purchase and install, but fluorescent bulbs are more energy-efficient, have a longer life, and can be dimmed with special equipment. However, as the lamp ages, fluorescent lights lose lumen output. The life expectancy of incandescent bulbs is usually 1,000 hours; fluorescent is up to 20,000 hours.

Small flock producer Robert Plamondon in Oregon has the following recommendations for small growers. He uses incandescent lighting because he believes it holds up better than fluorescent under free-range house conditions. Use a 60-watt incandescent bulb for every 200 square feet of henhouse Use flat reflectors to maximize light Clean/dust bulbs regularly Position light fixtures so people or birds don't run into them. Use a guard over the light to prevent breakage or simply suspend a bare bulb that will swing if hit Put a dab of petroleum jelly on the threads of the bulb to keep the contacts from corroding and to keep mites and other tiny bugs out of the light sockets. If you use fluorescent, use the kind with sealed ballasts, since the vents in vented ballasts let in moisture and dust. These are available from farm-supply businesses Use a timer to control the light period, because if you forget to turn the lights on, it can cause hens to stop laying. Use a wired-in electromechanical timer (rather than plug-in type), and check it regularly, resetting it after any power outage.

Use waterproof sockets: porcelain or plastic (not brass shell). Use heavy duty extension cords for portable houses and cover connections to protect them from the weather. Adding the lights in the morning instead of the evening will allow a natural dusk for the birds and allow them to choose their roosts for the night, or a dimmer can be used to create dusk conditions. On a timer, it is necessary to make adjustments (usually weekly) to keep the day length at a certain length. Instead of regularly adjusting timers, Plamondon uses a dusk-to-dawn lightsensing switch which will turn the light off during the day (no waste of electricity) and turn on at night.

He turns lights off in the spring. In a house without electricity, batteries can be used to power lighting, such as a 12-volt battery. An inverter can be used if there is a need to switch from DC to AC voltage. Companies in the U.K. offer 12-volt lighting systems particularly for mobile poultry housing. A solar panel can recharge batteries. Large Amish poultry houses without electricity are sometimes lighted with Coleman lanterns (which burn naptha gas) and kerosene lamps.

Litter

Litter management is very important in most poultry production systems. Floors in poultry houses are usually concrete, wood, or earthen, and litter is used to cover the floor. Litter dilutes manure and absorbs moisture, provides cushioning and insulation for the birds, and captures nutrients for spreading where desired outside. Litter is also a medium for birds to scratch and is important for welfare. Birds are also raised on slat flooring through which the droppings fall into a pit below and are later removed. Keeping droppings dry will reduce odours and flies. Common litter materials include soft wood shavings or rice hulls. Other materials that may be suitable include sand, recycled newspaper (no glossy or colored inks for organic production), dried wood fibre, peanut hulls, and chopped pine straw.

Small-scale poultry producers have tried out various materials and have identified some problems: hay and straw become slimy, chicks eat sawdust, wood chips are costly, and hardwood shavings can splinter and cause skin punctures. Litter material should be high in carbon to prevent loss of nitrogen and should compost well. Litter is normally spread two to four inches deep and maintained at 20 to 30 percent moisture. Birds have a concentrated form of waste called uric acid, which makes it possible to keep a lot of birds on litter, but moisture can build up.

If litter feels damp to the back of the hand, it is probably at least 30 percent moisture. The house should be ventilated well to remove moisture in the air, and water leaks or sources of moisture such as condensation from n-insulated metal roofs should be avoided. High moisture in litter is very problematic, resulting in cake or a nonabsorbent crust. Caking especially occurs under waterers or other high impact areas. Wet litter causes breast blisters and sores on the birds' foot pads and hocks, and pathogens and parasites such as coccidia proliferate. In wet litter, uric acid is converted by bacteria to ammonia. Ammonia is a toxic gas that can damage the respiratory system of the birds and make them more susceptible to infections.

Ammonia levels should not exceed 25 parts per million (ppm) in the house. Levels can be measured with an ammonia meter, but these are expensive. Inexpensive methods are ammonia strips, available from Micro Essential Lab, or dräger tubes, available from Fisher Scientific. Ammonia measurements should be taken at bird level on a regular basis and particularly at finishing. Fly larvae also grow in wet litter and can be a nuisance.

Although conditions should be maintained to prevent cake, some producers rototill their litter during production, while the birds are present, to loosen cake. After loosening, cake should be removed. Tilling the litter may cause a spike in ammonia that should be dissipated as quickly as possible through open windows or with fans. According to Virginia producer Joel Salatin, at a low stocking density, the bedding is tilled and aerated as fast as the birds manure and does not cake. Tossing in whole grains may encourage birds to scratch and till. Heavy broilers are not as active at tilling litter as layers.

Litter is removed after the flock is finished and the house cleaned. In meat bird production, the litter is often kept in place and reused for several flocks. If re-used, cake needs to be removed with a pitchfork or decaking equipment. Salatin has used pigs to break up cake at the end of a flock. The litter should be top-dressed with fresh litter. The litter should not be reused if disease occurred in the flock. Litter treatments are added in the conventional poultry industry to re-used litter to reduce the formation of ammonia from nitrogen by lowering the pH. Typical poultry litter has a pH between 9 and 10.

Ammonia release is low when the environment is acidic (pH is less than 7). Low pH will also inhibit microorganisms, including pathogenic bacteria like salmonella. Poultry Litter Treatment (PLT), or sodium bisulfite, is the most common litter treatment. Aluminum sulfate is also used to reduce ammonia release. These materials are not permitted in organic production.

Soft rock phosphate can be used as a litter amendment to control odour and to reduce flies. Hydrated lime is not permitted in organic production to deodorize animal wastes. Used litter is removed from large houses with machinery. In small houses, litter is removed by hand, which is very labour-intensive. After removal from the house, manure and litter are usually spread on pasture and other agricultural land. In many areas, poultry manure and litter are a great benefit and add valuable nutrients such as nitrogen (N), phosphorus (P), and postassium (K) to the soil. Poultry manure has 3.84 percent nitrogen, 2.01 percent phosphorus, and 1.42 percent potassium on a dry basis. On a fresh basis, there is more moisture in the manure, which dilutes the amounts of nutrients. As a rule of thumb, the amount of manure is equal to the amount of feed provided. Composting the litter adds further value to the manure because compost is an excellent soil amendment. More carbon material usually needs to be added to increase the carbon to nitrogen ratio. During composting, ammonia is released to the atmosphere, which lowers the nitrogen in the final product. Organic standards for compost require that starting carbon to nitrogen ratios be between 25:1 and 40:1 for information on composting poultry litter. Litter can also be composted in the poultry house after the birds have been removed. Windrows are made in the house and the litter is respread after composting. The building must be ventilated so that gases can escape.

The manure/litter from poultry houses has a natural tie to organic crop production. Synthetic fertilizers are not allowed in organic crop production; poultry litter has the advantage of being a natural fertilizer (as long as synthetic materials are not added to it). Litter from birds fed arsenic compounds is not permitted in organic production. Unfortunately, in high poultry-producing areas, manure/litter has become a liability because there is so much of it. Phosphorus is a nutrient pollutant because it may end up in runoff water, allowing algae to grow and contributing to water quality problems. Litter/manure cannot be dumped on land without consideration of crop/forage needs. Nutrient application from animal waste is becoming more regulated in the U.S., and nutrient inventories are kept on the farm. Regulations vary by state and are phosphorus-based or nitrogen-based. In European countries where there is little land compared to dense human populations, the regulation is nitrogen-based and is limited to 170 kg of nitrogen per hectare per year (equivalent to 149.6 lbs of nitrogen per acre). Best management practices are important in applying animal waste to land, such as incorporating litter instead of surface application, vegetative buffer strips to capture runoff nutrients prior to reaching waterways, etc. In parts of the U.S., since litter should not be spread during the winter because the ground is frozen or crops may only be fertilized in spring and summer, proper storage is required for litter. There are useful Extension publications on poultry litter application.

An alternative type of litter management is composting litter while the birds are in the house in order to reduce the volume of litter and create a healthy environment. This process, called "composting litter," has received little scientific attention since the 1950s. It usually starts with at least six inches of litter. The poultry till and aerate the litter or the litter may be tilled with machinery. Thin layers of fresh litter are added with new flocks or if the litter becomes wet. Small flock producer Robert Plamondon uses this technique and removes only half the litter at a time, when the accumulation becomes too much for the house.

Although composting litter is a form of composting or decomposition, it is not as efficient as the composting process described. The amount of decomposition that occurs depends on the amount of birds in relation to amount of litter and temperature. The carbon to nitrogen ratio is not likely to be ideal unless a lot of extra litter is added. Producer Joel Salatin adds enough litter to keep the carbon to nitrogen ratio at 30:1, but it is expensive. There may be some heat from decomposition, and ammonia gas is produced, so the house should be well ventilated. Salatin says the bedding pack must be at least 12 inches deep to work. Composting litter is rich in vitamin B12, most likely due to the presence of microbes. Some poultry producers are interested in beneficial microbes that may be present in composting litter to help induce immunity in birds, particularly during brooding. In fact, inoculating composting litter with microbes can facilitate management of the litter as a biologically active, organic substrate in a slow process of decomposition. Effective Microorganisms (EM), biodynamic preparations, and compost teas have been added to poultry litter to provide a healthy, probiotic environment, enhance bioprocessing, reduce ammonia, and reduce litter volume.

Air Quality

A poultry house of any size can have poor air quality if ammonia and dust levels are high. In large houses, air emissions to the outside are an issue for environmental air quality. Tree shelter belts have been used around houses as a way to capture emissions. Again, keeping litter dry helps reduce ammonia. In addition to ammonia and dust levels, it is also important to monitor hydrogen sulfide, carbon dioxide, and carbon monoxide levels, especially in large houses. Free-range systems have the advantage of fresh air.

Brooding Environment and Management

Brooding is a Critical Period for Poultry

New chicks can't maintain their own temperatures, so they are usually brooded until they are fully feathered. In natural brooding, the mother hen provides heat. In artificial brooding, heat is provided by an external heater. Large-scale broiler producers usually brood in the same building where the birds will be kept to market age, which is "one-stage production." Layers and turkeys usually use a "two-stage" system in which a different growing facility is used after brooding.

Small producers often use two-stage production; they brood in a central building and then move the birds out to small portable houses on pasture after brooding. The brooding building may be located close to home so the producer can keep a close eye on the young chicks. However, moving birds to a new facility after brooding is labour-intensive and a source of stress for the chicks. Brooding in the field in small portable houses reduces the need for moving chicks and allows early access to range, but field brooding requires an insulated house, small brooders, propane tanks, and battery run lights in each house.

Brooding can be "spot-brooding" vs. "whole house" brooding. Spot-brooding heats a localized area, while whole house brooding heats an entire room. This is also called "cool-room" vs. "warm-room" brooding, respectively. In the past, very cold-room brooding was practiced in drafty barns or other out-buildings. The brooder was surrounded with curtains or insulation to prevent heat from being lost to the room. It fell out of practice as brooding and poultry production moved to large poultry houses, and brooders of this type are no longer available. Robert Plamondon has been a leader in the U.S. in providing information to small-scale producers who use outbuildings and need good cold-room brooding technologies, including homemade insulated brooders. A brooding house must have good ventilation while preventing drafts. Although some small producers use a dedicated purposebuilt structure or building, many improvise with an outbuilding. Brooder guards (usually cardboard) stop floor drafts.

Types of brooders include:

- Heat lamps
- Hovers
- Space heaters
- Battery brooders.

Heat Lamps: Many small poultry producers use spot brooding in a variety of setups with an electrical heat lamp. Heat lamps are generally used above a box that keeps the chicks close to the heat source and reduces drafts. This setup is usually placed in a residence or an outbuilding. According to Plamondon, a 250-watt heat lamp suspended 18-24 in. Over the brooding area, completely surrounded by a draft guard 18-24 in. High, will brood 75 chicks at 50° F minimum room temperature. This method is dependent on the presence of an effective draft guard. Many hardware stores carry heat lamps. Rocking T Ranch and Poultry Farm maintains a website with information on homemade brooders.

Hovers: Hovers are brooders with a canopy to keep warm air close to the ground to warm chicks. Hovers are usually suspended from the ceiling. In large-scale production, hover brooders often have an umbrella or pancake shape and are fueled by propane or natural gas. Again, cardboard brooder guards provide protection from floor drafts. Farmtek carries propane brooders in various sizes. Standing hovers are placed on the floor above the birds. Plamondon describes a box-shaped, standing hover that is insulated and heated by electric lamps. It was developed in the 1940s by the Ohio Experiment Station and was popular for small flocks. It was designed for brooding under farm conditions—in drafty barns and in portable houses on range. It can be insulated by adding litter on top or by adding aluminized bubble wrap (Tekfoil©) inside. Brooder boxes are a type of very small hover. These are individual boxes that contain their own electrical heating element, feeder, and waterer and are placed on the floor on litter.

Space Heaters: Space heaters heat an entire area; they are not placed directly above the birds as hovers are.

Battery Brooders: Battery brooders are basically a unit of brooder boxes stacked on top of each other, separated by wire floors.

Hatcheries, such as Murray McMurray (13) and GQF (14) sell both brooder boxes and battery brooders. It may be possible to find older used battery brooders. When using a battery brooder, the room should be kept above 60° F and ventilated well. (5) Paper or plastic liners on the manure trays make removing manure easier. Propane usually keeps litter drier than electric heat lamps. It is easier to keep a stable temperature during brooding with propane heat and the addition of thermostats. Backup heat is needed for electrical set-ups, because electrical outages are always a concern. In the past, brooders were fueled by other means: kerosene, coal, and wood. The brooding area should be prepared with fresh litter and heated before the chicks arrive so that the litter is warm. The temperature at the start of brooding is 90°F and is reduced by 5°F every week for two to four weeks. The chicks should be able to move away from heat. The chicks are well distributed if the temperature is right for them. If it is cold, they will huddle. If it is too hot, they will spread away from the heat source. Chick-size waterers and feeders are used during brooding, because chicks can fall into waterers and get chilled. Feed should be provided on the floor in a shallow pan so the chicks can easily find it. Dipping the beaks of a few of the chicks in water and feed will help them learn quickly to eat and drink, and the other chicks will imitate them. When placing the chicks in the brooder, provide 24-hours of light to help chicks find food. Dark periods can be added after a few days. "Starve-outs" are chicks that don't learn to drink and eat. Turkey poults, in particular, are susceptible to this, as well as to stress and chilling. Electrolyte supplements can be added to water if chicks have been stressed during shipping. Sugar is also useful to provide energy.

According to Plamondon, the amount is one pound of sugar per gallon, which is about the same sweetness as Kool-Aid. Other producers have used one tablespoon apple cider vinegar and one teaspoon blackstrap molasses to a gallon of water. Supplements in the water are only useful the first day, because the chicks find the feed after that. Supplements need to be organic if the flock is certified organic. Some producers provide outdoor access or harvested forage during brooding. Finely chopped grass or pieces of turf may help chicks become accustomed to digesting forage and the microbes they encounter outside; sand or small grit should be provided to help their gizzards grind fibrous feed. Three-week-old chicks are still at risk from cold temperature when placed on pasture. They should have access to a warm place. Many producers use the insulated hover popularized by Plamondon in the field with older chicks. Even if only heated by the body warmth of the chicks, it can help protect chicks. The traditional time to move birds out to pasture is when they are wellfeathered. Fast-growing broilers usually leave the brooder at three weeks. Layer chicks are slower-growing and may need heat until four to five weeks, depending on weather.

General Management

In addition to providing the proper temperature, ventilation, lighting, and litter conditions for the birds, feeding, watering, and health are important parts of management. Birds and equipment should be inspected at least twice per day to monitor health and identify any problems. Caretakers should be trained in bird management and welfare. They should treat the birds calmly with no rough handling. Although birds usually have outdoor access, additional enrichments are used in houses to improve welfare of birds and include roosts, straw bales, and scratch grains.

Rodent Control

Rodent control is a very important management practice in poultry production; rats kill chicks, eat feed, and spread disease. Problem rodents for U.S. poultry production include the Norway rat (*Rattus norvegicus*), the roof rat (*Rattus rattus*), and the house mouse (*Mus musculus*). Rodent control is a systems-based approach of prevention and exclusion including the following.

Habitat Reduction: Vegetation should be kept short around houses, spilled feed cleaned up, hide-outs dismantled, including scrap piles. Rodents should be exposed so they are vulnerable to predation.

Exclusion: Concrete or gravel floors help keep rodents from tunneling into a house. In a small portable house with a raised floor, the space between the ground and the floor provides attractive, darkened nesting sites, unless the floor is high enough above the ground that rodents do not feel protected. Poultry feed should be stored in rodentproof containers.

Traps: Traps include snap traps, sticky traps, or mechanical "tin cat" traps. Predators. Cats and dogs can help control rodents; rat terriers are especially helpful. Barn owls eat rodents but also eat chicks unless the chicks are in predator-proof houses at night.

Bait: Rodenticides are chemicals that kill rodents and are incorporated into attractive food or baits. Most rodenticides are not permitted in organic programs. Anticoagulants. Since these rodenticides prevent blood clotting, the rodent dies through internal bleeding. The well-known Warfarin © was the first type developed. Multiple dose anticoagulants are the safest type to use, since a rodent has to nibble the bait several times to be affected. Single-dose coagulatants are more lethal and work faster but are less safe around children and pets.

Baits come in Several Forms: Blocks, bulk pellets, and pellet place packs. Norway rats live underground in burrows, where bulk pellets should be placed. Roof rats and mice can be controlled with blocks that are nailed or tied down to prevent them from dragging blocks away to store. Putting bait in a bait station will keep random animals from eating the bait. Baits are usually rotated to prevent rodents from becoming accustomed to them. Sulfur dioxide, or smoke bombs, are permitted in organic production as an underground rodent control. Deterrents. There are many types of deterrents such as sound or light. For example, owl predator lights will startle owls that are preying on birds at night. Some producers use blinking holiday lights. In addition to controlling rodents, wild birds should also be controlled because they can introduce disease. Screening openings will prevent birds from entering the house; nest should be removed from the house.

Family Poultry Production

The rearing of chickens is popular in rural areas of most resource poor countries, as a means of providing supplementary food, extra income and employment to family members and also to capitalise on harvest wastes and inferior grains produced on farms. The term "family poultry" refers to any genetic stock of poultry (unimproved or improved) raised extensively or semi–extensively in relatively small numbers. Most of family chicken production systems are based mainly on native, domestic species which require very low levels of inputs. The terms "indigenous or native chickens" are often used as a synonym to family chickens, even where there is often a high proportion of non-indigenous blood in the flocks. They are also termed "scavenger chickens" where they are allowed to run free in the yards or surrounds and "backyard chickens" where they are kept in a house yard (confined or free). Family chickens comprise the major part of the poultry industry in many developing countries. Family poultry production has been a traditional component of small farms throughout the developing world. In Africa it is estimated that 80% of the poultry population is found in these production systems, that contribute up to 90% of the chickens reared and they supply the bulk of the national requirements of eggs and meat for the urban populations. Attempts are being made to raise the productivity of family chickens in developing countries, by improving housing, nutrition and health programmes. Improvements in performance resulting from improved management (nutrition, housing and disease control) and marketing strategies have been reported in some countries such as Indonesia. Future prospects for rearing family chickens are believed to be good, because of traditionally high demand for their meat, which is tasty compared to that of commercial broilers. Family chickens are among the most adaptable domesticated animal species in all climatic conditions. They survive under unfavorable weather conditions, sheltered or not sheltered, in cages or in tree branches. The management is largely the responsibility of women and children. This manuscript presents a review of worldwide family poultry production including information from Botswana, and some African and Asian countries.

Multiple Roles of Family Chickens

The rearing of family chickens is most prevalent in rural areas where the cash incomes of the people are generally lower than in urban areas. In such areas, unemployment is often high and female labour is relatively underutilised so poultry keeping can help to supplement incomes and the nutritional status of families. Rural families sell some birds when the need for cash arises. Birds are also slaughtered to honour a friend or relative who has been away for some time, or taken as provision while traveling. Surveys on the economic importance of family chickens in South America, Asia and Africa revealed priorities for rearing as social/security (82%), consumption (24%) and sale (18%). In Botswana, it is estimated that priorities for rearing are: family consumption (95%), sale/source of income (65%), greeting visitors (55%), hobby (14%) and others (including healing rituals) (12%). The fact that family consumption and source of income ranked highest clearly indicates that family poultry plays an important role in poverty alleviation. Family chickens play an important role in traditional healing/curing rituals. For instance, in the Mandara tribe of north Cameroon, the strains with white plumage (Dzape) and the black ones (Dongwe) are relatively expensive, as they are used in traditional medicine and magical practices. White and black chickens are also used in healing rituals in Botswana. Previous study in Serowe-Palapye Subdistrict (Botswana) showed that a mixture of herbs, blood, viscera and chicken meat is used to treat intestinal worms (tapeworms and roundworms) in children, epilepsy and male sterility.

In addition, family chickens provide sanitation by way of making use of spilt grains and hence help in cleaning the yards. Some indirect benefits of rearing poultry include waste disposal, converting energy in left-over grains into valuable protein, utilization of kitchen scraps such as rice, bread and other foodstuffs, and chicken manure as a fertilizer used in gardens and orchards. The increased production of fruit and vegetables gives rearers additional income and directly contributes to the economy.

It is estimated that 15 chickens produce about 1-1.2 to 1.9 kg of manure per day. The input of organic manure encourages the development of earthworms in garden soil, which together with termites and insects, can later provide an additional source of feed to chickens. The organic content of manure helps maintain soil structure, and hence improve water retention capacity of the soil. However, poultry manure tends to acidify the soil and is usually deficient in potash. Consequently, soils fertilised with chicken manure should be dressed regularly with lime at a rate of 1 kg/m2. Chickens control weeds and insect pests by foraging. The foraging habits of the fowl make it a 'biological environment controller' that recycles waste and control pests and harmful insects. Therefore, chickens can be regarded as an important component in integrated farming system.

Advantage of Family Poultry Over Commercial Poultry

The advantages of rearing family chickens in the rural areas over the raising of industrial chickens include:

- ease of rearing;
- low input requirement;
- no need for permanent and/or expensive housing;
- possible low incidence of diseases;
- very stable price of finished products;

- market price is higher than that of broilers;
- free choice of rearers in time of selling chickens;
- scavenging and taking in natural feed, *e.g.*, grass and insects, and;
- chickens are kept to produce eggs and chicks for the next rearing.

A major disadvantage of family chickens is that if they are not confined they can be destructive to gardens and vegetable seedbeds. Family chickens may also be a nuisance to neighbours who grow flowers and vegetables leading to quarrels.

Characteristics of Family Chickens

Family chickens are generally of small body size, having slow growth rate with different colours of plumage, and of dual purpose type with variable body conformation and physical characteristics. Body weight is variable and they (family chickens) lack uniformity in growth. Family chickens are active, lively and fond of fighting (aggressive), especially when intensively reared. The hens have the instinct of broodiness. The family chicken has a single comb, black and grey shank. Meat from family chickens is often referred to as "delicious" and is a favourite in many developing countries. For this reason, they are sought at markets for their tasty meat, especially for ceremonies. The meat is tasty, relatively dry and well adapted to the prolonged African way of cooking. Their carcasses have less fat than commercial broilers. In many countries such as The Gambia, Indonesia, Kenya, Malaysia and Vietnam, free range family chicken meat and eggs fetch a higher price than commercial broilers. For instance, in Morocco the price of chickens is twice that of industrial broiler chickens. Eggs from family chickens are preferred because of deep yellow colour of the yolk.

Rearing Family Chickens

The extensive system of poultry keeping is considered to be quite efficient in rural areas. Capital and labour inputs are extremely low, so even though production levels are low, costs per kilogram of bird are very small. The family chicken is a relatively low producer because of its genetic make-up, low plane of nutrition and management in general. However, rearing is profitable mainly because the inputs required to sustain the system are small. Hence, it is often referred to as lowinput low-output system. In many instances rearers incur minimal to no cost because feeds are surplus or internally generated.

Management Practices

Family poultry production involves womenfolk and children more than menfolk. This is perhaps because men often work away from the house, growing crops or employed by someone else. The main characteristics of family poultry rearing include:

- free ranging during the day and gathering at night into a basic shelter to avoid loss through predation;
- feeding is mainly limited to insects, seeds and kitchen wastes (sometimes a supplement is provided but this depends on the availability of feedstuffs).
- very low productivity (the eggs are rarely harvested but rather the hens are allowed to brood them);
- flock numbers vary markedly, because of prevalence of diseases; and
- they are considered hardier, more attentive to dangers (predators and strange objects) than improved strains, and resistant to diseases, being better adapted to the local environment than improved commercial breeds. However, disease resistance is a contentious claim that deserves empirical verification.

Housing

Generally, scavenging chickens are not housed except possibly at night, because they must be allowed to find their own food at minimal cost. As a result, birds sleep on trees, piles of bricks/blocks, old vehicles, bush fences, walls, under roof overhangs or on top of the huts, thus exposing themselves to the risks of predation, climatic hazards and theft. Minimal housing is provided at night, for protection from predators, and an enclosure of some type for part of the day is used to facilitate egg collection.

The risk of predation and theft is common with birds that are not confined at night than with those that are. When shelters are provided, they are often made of materials that are easily available such as old tins, iron sheets, plastic bags and thatch grass. Shelters are usually built at the back of the owner's houses/huts. The roof may be of grass thatch or galvanised iron sheets. Because of the nature of the housing system, predators, particularly cats, may cause losses in chicks. Both adult and young males are the major contributors in constructing shelters.

Feeds and Feeding

Family chickens usually have to find food for themselves. The opportunity to scavenge is a way of allowing the chickens to correct any nutritional deficiency in the feeds offered as supplements. The free range system makes it difficult to measure feed consumption, body weight and egg production. Although feeding is mainly limited to insects and kitchen wastes, bran (mainly sorghum) and whole grains are sometimes used as well. Bran is widely fed, especially to chicks, and is obtained from milling plants found in villages or is generated from homes. Bran is may be given wet or dry in various containers or on bare ground. Broken cooking pots, old automobile tyres (cult in half) and tin cans are the vessels used for feeding and drinking. A few rearers (about 1%) in Botswana feed their chickens on compounded diets (broiler starter, finisher, growing or laying diets).

Scavenging chickens start roaming the fields in the morning to search for feeds such as earthworms, beetles, spiders and scorpion, grasshoppers, centipedes, lizards, grass and legume seeds, berries, green leaves *etc.* and return to the farmer's house in the late afternoon. In addition to scavenging, birds are fed maize chaff, cowpea testa, melon fruits, kitchen waste etc. Birds are fed at different times of the day depending on feed and labour availability.

They are fed once, twice or three times a day and are fed mainly in the morning before they roam the village outskirts in search of feed and late afternoon to encourage them to return home. Once a day feeding is common during periods of feed shortage, especially in summer. At times they are not fed at all leaving them to depend entirely on scavenging. The common feeding method is by broadcasting grain on the bare ground. Generally, rural farmers only occasionally supplement feeds in small amounts when they can afford it. There is evidence that if feeding of family poultry is improved, they will be more productive. Improved feeding is particularly important if indigenous breeds are upgraded using exotic and more productive breeds. Feeding may be improved either by allowing the birds to scavenge and be fed a daily ration or they may be enclosed and fed complete diets. In Nigeria, it is reported that supplementation at 30 g/bird/day enhanced the growth rate of chicks while supplementation below 60 g/day was insufficient to increase the growth rate of adult birds.

Drinking Water Sources

Most rearers give chickens water used for human consumption. In the villages chickens are usually given borehole water, while at the fields (*masimo*) water from streams or wells are used. However, it is a common practice for rearers to give "wash-up water" containing food particles to their flocks. In most cases, chickens have to water from the nearby standpipe.

Several types of vessels are used as drinkers, including old metal (broken pots and lids of various containers) and plastic containers, troughs (metal and concrete) and old automobile tyres. Drinkers are placed in many places within the yard, usually under trees to keep water cool. To prevent containers from tipping, containers are half buried in the soil or a stone is placed in them. The containers are recharged once or twice a day. Some chickens are not provided with water for the whole day while out scavenging, which could represent a serious constraint to production in some villages and seasons.

Breeding in Family Chickens

Family poultry rearers worldwide do not follow a planned breeding programme, with the results that close inbreeding often occurs among indigenous flocks. Males or females are usually allowed to run together all the time. The ages and numbers of young birds fluctuate markedly because the hens tend to breed all year round. In Botswana, the most prolific breeding time for family chickens appears to be in autumn and winter because of low incidence of Newcastle disease (ND), low predation rates, low parasite populations and an abundance of feed supplies. Natural brooding produces most chicks of indigenous breeds. The hen usually lays eggs in the nest she makes in the bush, bush fence, or in the nest provided.

Eggs found laid in the bush are collected and placed in the nest prepared by the owner in the yard. The eggs are naturally hatched by family hens, which consequently remain broody (and do not lay) for periods of the year. Fertility and hatchability percentages are reported to be 80-82%. The chicks are brooded by the hen, which is usually very aggressive and protective. A large number of chicks die from diseases or are killed by predators within the first two months. Mortality also occurs due to the hostile environment that the newly hatched chick confronts, *e.g.*, unfavourable weather conditions such as hail and storms. Other causes of mortality are bur-bristle grass (*bogoma*), accidents caused by vehicles and drowning, especially of chicks. Some birds get stolen while those that stray into neighbours' gardens or homes are likely to be killed. An average mortality of 60-68.5% for young chickens up to six weeks of age is common.

Productivity in Family Chickens

Family poultry production levels are low, eggs are small and mortality is high compared to commercial production. The lower egg weight correlates with their lower body weight. Egg weight seldom exceeds 42 g and egg production is below 100 per bird per annum. However, egg weight of 45-50 g has been reported in some African countries. Sexual maturity in family chickens is reached at about 6-8 months. The length of time it takes a chick to mature depends mainly on feed availability. There is a wide range of values reported on the number of eggs produced by the hen in a year. Family hens lay between 30 and 144 eggs per year. About 2.6 to 5.2 clutches are produced in a year. In Botswana, 3 to 4 clutches are produced per year and clutch size of 14-20 eggs is produced. Family hens can stay with the brood for up to 2-3 months, by which time the young growers had separated themselves. The hen then commences to lay another clutch with an average interval of 9.3 weeks. The owner consumes a few eggs while the hen hatches the rest. The family chicken's growth rate and feed consumption rates are not as high as exotic broiler breed. About 3.03 kg of feed is needed for each kilogram grain at 14 weeks of age compared with less than 2 kg feed/kg of live weight for broilers. Market weight of 1-1.5 kg is reached in 4-5 months.

Factors Affecting Productivity

Factors that reduce productivity include:

- lack of technical assistance from extension services;
- poor housing and feeding of birds;
- poor breeding stock;
- poor disease control (notably ND);
- climatic hazards; and
- predation by snakes and birds of prey.

Low productivity may also be due to the genetic make up of family chickens, inherent and pronounced instinct of broodiness of hens and a long period of nursing baby chicks.

Health Control

Diseases and Control

The rearers seldom carry out disease control measures because of relative unavailability of reliable vaccines and the high cost of medications or vaccines. Vaccination coverage among family chickens was estimated at only 20% in Indonesia. Previous study showed that in Botswana, only 2% of family chicken rearers use conventional vaccines. The belief that family chickens are less susceptible to diseases could be contributing to lack of disease control. Family chickens are considered to be a major reservoir of infections for commercial chickens. Disease control involves modern and traditional remedies and the latter predominates. Several studies have shown diseases to be the major constraint to the development of family poultry, notably ND. Newcastle disease is the most serious endemic disease of poultry throughout the African continent and causes 70-100% mortality in family chickens, especially in young chickens. In the Democratic Republic of Congo, the rearers know ND as the "bomb" because it causes heavy mortality after its occurrence like a bombshell will do. Newcastle disease is endemic in Botswana with most of the outbreaks occurring in the warm months (August to December). The ND outbreak that occurred in 2005 resulted in over 6000 commercial broilers reported dead, while the number of family chickens that died is unknown. Other diseases include fowl pox, cholera, typhosis, coccidiosis, enteritis, avian mycoplasmosis, chronic respiratory disease, fowl typhoid and ectoparasitism. Several sulfonamides or broad-spectrum antibiotics are used in treatments. Family chicken rearers do not take any particular control measures. Sick chickens and their contacts are disposed of through markets or are slaughtered for consumption.

Conventional vaccines such as La Sota and Hitchner 1 are of limited value in family husbandry system, as the flocks are small, widely scattered, the birds are free-ranging, and young birds are constantly hatched and added to the flocks. These vaccines must be kept in cold chain and will lose their effectiveness if left out of the refrigerator for more than a few hours. In most cases refrigeration facilities are not available for protecting conventional ND vaccines. In addition, these vaccines are seldom available in the rural areas. As a consequence, the heat tolerant vaccines (Australian V4 and I-2) have been developed for this difficult environment where it is difficult to keep vaccines cold. Vaccines can be given by placing a drop in the eye, via drinking water and some by injection.

Bibliography

Andrews L.: *Citrus Production - Orange*, St. Augustine, Trinidad and Tobago, 1990

Bansil P.C. and Malhotra S.P. : *Livestock Economy of India*, CBS Pub, Delhi, 2006.

Batra Shikha : *Livestock in India : A Sociological Analysis*, Vista International Pub, Delhi, 2005.

Bhaskaran S and Suchitra Mohanty : *Marketing Of Livestock And Livestock Products In India*, Icfai University Press, Delhi, 2007.

Bogdan, A.V. : *Tropical Pasture and Fodder Legumes,* London: Longmans, 1977.

Butterworth, M.H. : *Beef cattle nutrition and tropical pasture,* London: Longman, 1985.

Das Kornel : *Tribal Crop-Livestock Systems in South-East India*, Manohar, Delhi, 2006.

Das N., Misra A.K. : *Forage for Sustainable Livestock Production*, Satish Serial Pub, Delhi, 2009.

Fairey, D.T., & Hampton, J.G. : *Forage Seed Production of Temperate Species,* Farnham Royal: CAB, 1997.

Frame, J. : *Improved Grassland Management,* Ipswich: Farming Press, 1992.

Gangadhar K.S. : *Livestock Economics : Marketing, Business Management and Accountancy*, New India Pub, Delhi, 2009.

Ghosh Bibek : *Trends and Issues of Livestock Production Systems*, Gene Tech Books, Delhi, 2007.

Harold C Long : *Plants Poisonous to Livestock*, Asiatic Pub, Delhi, 2006.

Hitchcock, A.S., & Chase, A. : *Manual of Grasses of the United States,* USDA, 1971.

Holmes, W. : *Grass, its production and utilisation,* Oxford & London: Blackwell, 1989.

Honnappagol Suresh S. : *Reproductive Disorders of Livestock : Prevention and Management*, Agrotech Pub, Delhi, 2006.

Khandelwal M K and Yadav Mukesh : *Potentialities and Prospects of Livestock Development*, RBSA Pub, Delhi, 2007.

Kundu S.S., Chaturvedi O.P. : *Environment, Agroforestry and Livestock Management*, International Book Distributing Co., Delhi, 2008.

Kurup M.P.G. : *Livestock in Orissa : The Socio-Economic Perspective*, Manohar, Delhi, 2003.

Mathialagan P.: *Textbook of Animal Husbandry and Livestock Extension*, International Book Distributing Co, 2005.

Mishra R. K: *Livestock-Crop Production Systems and Livelihood Development*, Atlantic, Delhi, 2007.

Mishra S.N. and Dikshit A.K. : *Environment and Livestock in India : With a Comparative Study of the Indian and US Dairy Systems*, Manohar, Delhi, 2004.

Murthy A.P. Narayana : *Sustainable Livestock Production in India*, Adhyayan, Delhi, 2009.

Pandey D N and Bajpai Amita : *Recent Trends in Animal Nutrition and Feed Technology for Livestock, Pets and Laboratory Animals*, International, Delhi, 2003.

Pathak P.S. and Kundu S.S. : *Livestock Feeding Strategies for Dry Regions*, International Book, Delhi, 2006.

Patil N.V., Mathur B.K. : *Feeding and Management of Livestock During Drought and Scarcity*, Scientific Pub, Delhi, 2010.

Randhawa, M.S. : *Farmers of India*, I.C.A.R., New Delhi, 1959.

Raymond, F. & Waltham, R. : *Forage conservation and Feeding,* Ipswich: Farming Press, 1986.

Ruhela Archana and Sinha Malini : *Livestock Economics*, Oxford Book Company, 2010.

Russell, Howard S. : *A Long, Deep Furrow: Three Centuries ofFarming in New England.* Hanover, N.H.: University Press of New England, 1976.

Sharma G.R.K. : *Cyber Livestock Communication and Extension Education*, Concept, Delhi, 2006.

Sharma M.C. , Tiwari Rupasi : *Entrepreneurship in Livestock and Agriculture*, CBS Pub, Delhi, 2010.

Singh Sanjeev Kumar : *Extension Techniques for Livestock Development*, New India Publishing, Delhi, 2011.

Singh, P. : *Forage Research - Present Status and Future Strategies: Pasture and Fodder Crop Research,* RMSI, IGFRI, Jhansi, 1988.

Snaydon, R.W. : *Managed Grasslands,* Amsterdam & London: Elsevier, 1987.

Tulachan Pradeep M : *Community Empowerment in Livestock Resource Planning : A Suggested Participatory Policy Framework*, International Centre for Integrated Mountain Development, 2002.

Walter H. Peters and Robert H. Grummer : *Livestock Production*, Greenworld, Delhi, 2001.

Walton, P.D. : *Production and Management of Cultivated Fodders,* Reston, VA: Reston Publishing, 1982 .

Wani Gazanfer and Wani Paras : *Reproductive Physiology and Livestock Embryo Transfer*, Satish Serial Publishing House, Delhi, 2010.

Index

❑❑❑